T0291009

ARTIFICIAL INTELLIGENCE-AIDED MATERIALS DESIGN

ARTIFICIAL INTELLIGENCE-AIDED MATERIALS DESIGN

AI-Algorithms and Case Studies on Alloys and Metallurgical Processes

Rajesh Jha and Bimal Kumar Jha

CRC Press
Taylor & Francis Group
Boca Raton London New York

CRC Press is an imprint of the
Taylor & Francis Group, an **informa** business

First edition published 2022
by CRC Press
6000 Broken Sound Parkway NW, Suite 300, Boca Raton, FL 33487-2742

and by CRC Press
2 Park Square, Milton Park, Abingdon, Oxon, OX14 4RN

© 2022 Taylor & Francis Group, LLC

CRC Press is an imprint of Taylor & Francis Group, LLC

Library of Congress Cataloguing-in-Publication Data
Names: Jha, Rajesh, author. I Jha, B. K. (Bimal K.), author.
Title: Artificial intelligence-aided materials design : AI-algorithms and case studies on alloys and metallurgical processes / Rajesh Jha and Bimal Kumar Jha.
Description: First edition. I Boca Raton, FL : CRC Press, 2022. I Includes bibliographical references and index. I Summary: "This book describes the application of artificial intelligence (AI)/machine learning (ML) concepts to develop predictive models that can be used to design alloy materials. Readers new to AI/ML algorithms can use the book as a starting point and use the included MATLAB and Python implementation of AI/ML algorithms through included case studies. Experienced AI/ML researchers who want to try new algorithms can use this book and study the case studies for reference. This book is written for materials scientists and metallurgists interested in the application of AI, ML, and data science in the development of new materials"– Provided by publisher.
Identifiers: LCCN 2021047601 (print) I LCCN 2021047602 (ebook) I ISBN 9780367765279 (hbk) I ISBN 9780367765286 (pbk) I ISBN 9781003167372 (ebk)
Subjects: LCSH: Metallurgy–Data processing. I Alloys–Data processing. I Artificial intelligence–Industrial applications.
Classification: LCC TA483 .J48 2022 (print) I LCC TA483 (ebook) I DDC 620.1/6–dc23/eng/20211117
LC record available at https://lccn.loc.gov/2021047601
LC ebook record available at https://lccn.loc.gov/2021047602

ISBN: 978-0-367-76527-9 (hbk)
ISBN: 978-0-367-76528-6 (pbk)
ISBN: 978-1-003-16737-2 (ebk)

DOI: 10.1201/9781003167372

Typeset in Times
by MPS Limited, Dehradun

Dedication

Essentially, all models are wrong, but some are useful.

George E. P. Box

A man would do nothing if he waited until he could do it so well that no one could find fault.

John Henry Newman

The most important thing is to find a problem and start working with the data in hand. Every algorithm, models and software will have flaws. New programming languages will be launched, and software will always have new versions. One cannot wait for the most perfect algorithm that will help them develop the most perfect model by using the most uniformly distributed data in the variable space. In the domain of metallurgical processes and materials design, one has to adapt and work with certain degree of uncertainty.

"What does your Master teach?" asked a visitor.

"Nothing…" said the disciple.

"Then why does he give discourses?"

"He only points the way; he teaches nothing"

One-minute wisdom

In this book, we have pointed ways by which a person can use the data generated through experiments, industries, simulations, or from the open literature and efficiently process it through a set of computational tools. In the case studies, our emphasis was on analyzing patterns in the data set along with developing accurate predictive models. We have presented case studies where we developed in-house code, used proprietary software, and performed calculations on a supercomputer. We have also listed a number of ways through which a user can perform lots of those complex tasks without having to write a single line of code, by working on a normal laptop/desktop, on a web browser, on a normal Android phone. The content of this book is in accordance with the Industry 4.0 and MGI/ICME initiatives and will be helpful in realizing the "Virtual Materials Design" paradigm.

Contents

Foreword

Artificial intelligence is a vast topic. A user needs to know which algorithm is suited for the specific problem in hand. Working on any particular popular AI-based approach may or may not work for a specific problem, just like a user follows different approaches for understanding the same metallurgical phenomenon. Industries may follow different approaches for manufacturing a similar product. For example, blast furnace iron making is one of the most efficient approaches among all iron-making routes. But there exist several other coal- and gas-based extraction techniques like Corex, Finex, DRI, HBI etc., that are in practice in different parts of the world. Similarly, the LD steelmaking process is the most common route of primary steel refining, while electric arc furnace is also widely in practice. From physical metallurgy point of view, nucleation and growth of a crystal/grain can be studied through mathematical modeling by Johnson-Mehl-Avrami-Kolmogorov (JMAK) or Kampmann-Wagner numerical (KWN) approach. Hall-Petch equation works well for grains above 100 nm, but the inverse Hall-Petch effect has been widely observed by researchers for nanocrystals. Thus, a user needs to be careful in choosing an AI-based approach for their problems just like they are careful while calibrating equipment before performing experiments or choosing a governing equation while performing mathematical modeling or following a particular approach in the industry for design and manufacture of a certain product.

This book will be quite helpful for understanding various aspects of artificial intelligence through several case studies within the domain of metallurgical and materials engineering. The authors have described several topics, while specifically stating pros and cons of different algorithms based on various concepts within the domain of artificial intelligence. Topics like alloy design has been discussed in detail. Two industrial case studies, on blast furnace iron making and LD steelmaking have been discussed in detail and are quite interesting from the industrial application point of view.

I am glad to write this "Foreword" for the book authored by Dr. Rajesh Jha and Dr. Bimal Kumar Jha. The authors have introduced a complex topic of artificial intelligence in a simple and understandable way through several case studies. This book will be beneficial for readers who want to take advantage of improving on their work by application of artificial intelligence algorithms. Several AI-based algorithms have been discussed and the authors have provided with resources that will be beneficial to readers. This book will provide a platform for readers who want to use AI-based approaches, but lack resources to start. Since all the work presented in this book was performed on a normal laptop computer, a user can just use their existing device and start with their current problem. The authors have provided with case studies where they have written their own computer program as well as case studies where they didn't write a single computer program but still were able to address their problems. This aspect of the book will be appealing to a lot of readers.

My compliments to Dr. Rajesh Jha and Dr. Bimal Kumar Jha for their commendable efforts in writing this book. It will be beneficial to readers in industry and research laboratories and can be used as a textbook in academic institutions. My best wishes for the success of the book.

Dr. Sanak Mishra, PhD
University of Illinois at Urbana-Champaign, USA FNAE, FNASc, FIIM, FIE, FCSI, FIOD, FAIMA, FDMA
Past President, Indian National Academy of Engineering
Formerly: Managing Director Rourkela Steel Plant and Member of the Board of Directors of Steel Authority of India Limited (SAIL)
Vice President of ArcelorMittal and CEO, India Projects
President, Indian Institute of Metals

Preface

There is a rapidly increasing demand for new materials to address the challenge faced by imposition of various environment norms which calls for clean energy from renewable resources, fuel-efficient automobiles or electric automobiles, efficient magnets, etc. Discovery of a new material or improving upon multiple properties of an existing material is a complex and time-consuming process. An experimentalist needs to test a new composition, design new manufacturing protocols for these new or existing materials, and perform characterization to understand correlations between composition, processing, microstructure, and desired properties. An industry or a funding agency demands optimization of multiple properties prior to deployment of these new proposed materials or manufacturing protocols. The traditional approach to new alloy design is based on the subjective experience of the designer, intuition, and is prohibitively costly. Random experimentation may take decades from discovery to deployment. Consequently, rapid progress in various computational tools development and application in materials design has evolved during the past two decades. The entire effort was aided in 2011 when the U.S. Congress approved a multi-million dollar "Materials Genome Initiative (MGI)" that focused on utilizing the expertise of professionals from academia, industry and research labs, and motivates them to use computational tools that will be helpful in accelerating the materials discovery process. Multi-objective optimization, statistical toolboxes, and machine learning approaches have been efficiently utilized over the years in addressing some of the challenges of the current state-of-the-art approaches in data-driven materials science.

A material scientist/metallurgist regularly deals with data from different sources including (but not limited to) data from different types of experiments in the lab, imaging, computer simulations, or data from industry. One of the main challenges faced by a materials design engineer is the lack of literature on choosing the correct algorithms for dealing with different types of data. It will become clear from the case studies presented in this book that a materials scientist/metallurgist may have to deal with different types of data from different sources and in different format. Thus, choosing the most appropriate algorithms is essential in the complicated task of developing predictive models for a given type of data. For example, data from industries such as an operating iron-making blast furnace is extremely noisy. Most of the time, there will be insufficient data and it will contain unreliable data that cannot be avoided as sensors pick up data at very high temperatures. Data from experiments can be highly non-linear by nature and will be difficult to analyze using statistical tools. Data from "CALculation of PHAse Diagram (CALPHAD)" approach comes from mathematical models, but still can be highly non-linear as often one has to deal with metastable phases while dealing with multi-component systems. Thus, one has to deal with missing data while working with data from CALPHAD approach. Also, data sets can have multiple design variables like operating parameters for an experiment or industrial process or chemistry in case of a material/alloy design. Similarly, there can be multiple conflicting objectives to be optimized simultaneously, such as productivity of a furnace and CO_2 content in top gas from a furnace or optimization of several mechanical properties of a material/ alloy. Dealing with multiple design variables and multiple simultaneous objectives on different types of data is a challenging task.

Thus, there is a need for a comprehensive guidebook for materials science/ metallurgy community where a researcher can find information on dealing with different types of data, several concepts of artificial intelligence that can be helpful in analyzing the data, advantages and limitations of each of the statistical and machine learning approaches, and can learn about pertinent computational tools to start the design process.

The purpose of writing this book is to provide the readers with a comprehensive guidebook that is easy to follow for researchers who want to efficiently generate and utilize their data by providing them with vital information on materials/ metallurgical concepts, advantages and limitations of several concepts of artificial intelligence or machine learning algorithms, along with several case studies along with their implementation in MATLAB® and Python languages. This book will cover ways on how to deal with data generated through the CALPHAD approach in detail as it is an important computational tool for material scientists and metallurgists. Readers can further find information on using machine learning algorithms in their own work by writing their own computer codes or using resources where a user can directly access AI/ML algorithms without writing their own code. All of the case studies were developed on a normal laptop/PC; thus, a user can follow this book on a normal computer and no investment is needed for accessing supercomputing resources. We have also provided ways to use some of these case studies on a common Android mobile device.

This book will also be beneficial for readers experienced in AI/ML. In the AI/ML community, there is a habit of following approaches by certain groups. For example, currently there are several groups that extensively use the Random Forest algorithm for developing predictive models, while some are solely working on Deep Learning approach using the TensorFlow/Keras library in Python. Similarly, in the optimization community, researchers prefer a certain Evolutionary Algorithm (EA) or Genetic Algorithm (GA) approach. In real-world problems, overdependence on a certain approach can result in not fully optimized results and even in unreliable results. We have demonstrated it in detail in this book through the case studies on problems dealing with different types of data. Data used in this work for developing case studies include experimental data, imaging data, data from industry, and data from simulations. Renowned statistician G.E.P. Box once quoted, "Essentially all models are wrong, but some are useful". Thus, an experienced user will also get useful information on algorithms that are best suited for a given type of data.

In the last few years, several start-ups have evolved and are offering their AI-based services at a given cost. Through this book, a researcher can better communicate with an AI expert in explaining their problem and expressing their concerns while applying the AI-based solution suggested by the expert. This is important as lots of materials and their manufacturing protocol are time tested and serve the need of the hour. Any solution offered by an AI-based approach must address all the concerns of an experimentalist, be it in designing a new composition or in designing a new manufacturing protocol.

This book covers various aspects of application of artificial intelligence algorithms to real-world problems and includes examples where a reader does not need to write a code and can work on a normal laptop/desktop and even an Apple IOS or an Android cell phone. Thus, this book can be used to understand applications of AI-based algorithms within the domain of metallurgy/materials science, motivate researchers to start working on AI-based algorithms in their group, as well as be helpful in communicating with an AI professional if they are planning to invest resources while choosing services offered by a professional. Case studies in this book are in accordance with the guidelines set by Industry 4.0 initiative.

Acknowledgments

First of all, we would like to thank the editors, Ms. Allison Shatkin and Ms. Gabrielle Vernachio, for their help and assistance throughout the process of book writing.

We would like to express our gratitude to Dr. Sanak Mishra for writing a Foreword of our book.

We would like to thank our family members for their moral support.

Finally, we are thankful to members of the team at CRC Press, Taylor & Francis Group for preparing the cover, typesetting, and copyediting of our book.

Authors Biographies

 **Dr. Rajesh Jha** is a postdoctoral researcher at the Department of Mechanical and Materials Engineering, Florida International University (FIU). Prior to FIU, he worked as a postdoctoral researcher at the Department of Mechanical Engineering, Colorado School of Mines, Golden, Colorado. He graduated with a Ph.D. in materials science and engineering from Florida International University, Miami, FL in 2016; master of technology (M.Tech.) in metallurgical and materials engineering from Indian Institute of Technology, Kharagpur, India in 2012; and B.Sc.(Engineering) in metallurgical engineering from BIT, Sindri, Jharkhand, India in 2009. Dr. Jha has worked as a faculty in Metallurgy at OP Jindal University, India. He has a strong background as an experimentalist and has worked as a Project Assistant at National Metallurgical Laboratory, Jamshedpur, India, working on microscopy, additive manufacturing, coatings, and corrosion. Since 2010, he has been working extensively on machine learning (ML)/artificial intelligence (AI) algorithms for developing prediction models, along with evolutionary (EA) or genetic algorithms (GA) for multi-objective optimization. He has applied AI/ML algorithms in the field of materials and alloy design: advanced high-strength steel (AHSS), soft magnetic (FINEMET) and hard magnetic (AlNiCo) alloys, nickel-based superalloys, titanium alloys, aluminum alloys, CALPHAD, process metallurgy: blast furnace iron making, LD steel making, nano-tribology, and image processing. He delivered a lecture on AI/ML-based algorithms application in alloy design at European Materials Research Society (E-MRS) Fall 2021. He also delivered a lecture/webinar on "Artificial Intelligence Aided Materials Design" on the invitation of ASM International East India Chapter in December 2021. In March 2021, he received the best poster award for his AI/ML based work at US NSF-JST (Japan) joint workshop on Materials Informatics and Quantum Computing. He has published over 30 publications, including two book chapters, journal articles, and peer reviewed international conference proceedings. He has developed a software for simulating nanoindentation through ML and is currently working on patenting/licensing his software. He has served as an editor and on the reviewer board of academic journals and has reviewed articles for 19 international journals on multi-disciplinary topics. Dr. Jha received appreciation for his role as an outstanding reviewer for the month of September 2021 from Transactions of Indian Institute of Metals, Springer-Nature. Dr. Rajesh Jha has worked in India, USA and Finland at various universities and research organizations. He has collaborated with researchers in India, Finland, Brazil, Italy, Russia and USA on multi-disciplinary topics.

Dr. Bimal Kumar Jha, Former Executive Director in charge, Research & Development Centre for Iron & Steel, SAIL, Ranchi, Former Visiting Professor at IIT, Roorkee, Professor, NIFFT, Ranchi

Dr. Bimal Kumar Jha, former executive director in charge of the Research and Development Centre for Iron & Steel, Steel Authority of India Limited (SAIL), Ranchi, graduated in metallurgical engineering from University of Roorkee in 1978 with the unique distinction of being awarded with all five medals of the Metallurgical Engineering Department. He joined RDCIS, SAIL in 1980 after completing a M.Tech at IIT Kanpur. Later on, he completed his Ph.D. on TRIP Steels from the University of Roorkee in 1996 as an External Candidate.

Dr. Jha, in his various capacities in RDCIS, has spearheaded the product development and its commercialization activities of SAIL from 2005 to 2015. He has been instrumental in the development of a number of important steel grades such as high-strength LPG steel for lighter domestic cylinder; seismic resistant TMT rebars; plates and structural; high-strength roof/rock bolt quality TMT rebars; advanced steel for the automotive segment like dual phase, super EDD, low-carbon boron, high-strength formable quality, SAIL FORMING, Mn-B Steels, etc. Dr. Jha, as head of RDCIS, made outstanding contributions towards formulation of the company-wide R&D Master Plan. Under his leadership, a process was evolved to identify and capture R&D expenditure in SAIL plants/units, which resulted in its increase from Rs 133 crores in 13–14 (0.24% of sales turnover to Rs 290 crores in 15–16 (0.58% of sales turnover)). He made immense contributions for preparation of the technology vision document 2025 for SAIL.

His keen interest in research and development is amply manifested through more than 140 publications in journals of national and international repute and the filing of 45 patents to his credit. Impressed with his seminal research work, ASM International and Taylor & Francis, USA conferred him the honor of contributing chapters on "Heat Treatment of Boron Steels" in the *ASM Handbook* and "Forming Behavior of Boron Added Low Carbon Unalloyed Steel" in the *Encyclopedia for Iron, Steel and their Alloys*. Dr. Jha is a widely traveled person. He has visited TSINICHERMET, USSR on a training program; university of California, Berkley, under the INDO-US NSF program; CBMM Brazil under the CBMM-SAIL collaborative program; and Belgium, Holland, France, and Germany under a management development program.

He has been involved in setting up a national platform SRTMI (Steel Research and Technology Mission of India), a Ministry of Steel, Govt. of India. Also, as a Steel Industry-Academia Interface (SIAI) convenor, he has been pursuing activities related to the formation of SRTMI and Operating Committees on R&D to foster steel research and major technology developments of national importance in the steel sector.

Dr. Bimal Kumar Jha has been conferred the "National Metallurgist Award (Industry)" by the Ministry of Steel, India, the highest honor in the country for the metallurgical profession, in 2015. He has also received the prestigious "Metallurgist of the Year Award" in 2002 from the Ministry of Steel and O.P. Jindal Gold Medal in 2014 from the Indian Institute of Metals (IIM). He was visiting professor at IIT, Roorkee from 2018–2019 and professor at NIFFT from 2019–2021.

1 Introduction: Artificial Intelligence and Materials Design

Today the computer is just as important a tool for chemists as the test tube. Simulations are so realistic that they predict the outcome of traditional experiments.

Royal Swedish Academy in its announcement of the winners of 2013
Chemistry Nobel Prize (NobelPrize.org 2013)

1.1 DATA-DRIVEN MATERIALS SCIENCE: INITIATIVES, GOALS, AND PROGRESS

Developing a new material/compound or improving properties of a known material is a complex task. Random experimentation can be time consuming and can cost a fortune (Morgan and Jacobs 2020, Mueller et al. 2016). In cases like drug discovery, random experimentation, and delay in the discovery of a cure can be catastrophic on a global scale, particularly during the current COVID crisis.

As stated by the Royal Swedish Academy, computer simulations can be helpful in simulating traditional experiments, but designing a material/compound includes handling several parameters at multiple length and time scales. Multi-scale materials modeling approaches include, but are not limited to, atomistic simulations, phase field modeling, CALculation of PHAse Diagram (CALPHAD), finite element analysis (FEA), etc. Designing a new material or a metallurgical process may involve all of these approaches. A major challenge is handling parameters for each of these approaches. Molecular dynamics (MD), ab-initio or density functional theory (DFT) can both be included under atomistic simulations but have different governing equations (Rickman et al. 2019). Additionally, output from one approach can be the input of another approach, like output from of atomistic simulations can be input for CALPHAD, cellular automata (CA), phase field, and even FEA models (Chakraborti 2004, Dimiduk et al. 2018, Mueller et al. 2016, Rickman et al. 2019). This makes the process extremely complex as uncertainty at any time and length scale can affect the outcome of the entire materials modeling process. Computer simulations can be helpful but atomistic, FEA, or CA-based simulations are computationally expensive and simulation time can vary from a few hours to a few days, to weeks, or even months. Additionally, a user needs to have access to supercomputers or high-performance computing (HPC) resources, which is again expensive to acquire and needs a team of experienced professionals to operate

DOI: 10.1201/9781003167372-1

(Dimiduk et al. 2018; Ghosh and Dimiduk 2011; Morgan and Jacobs 2020;Mueller et al. 2016).

Mathematical models or surrogate models can be effective in determining various correlations among a set of input and output parameters. Statistical tools can prove to be helpful in the development of these surrogate models, but statistical toolboxes have its limitations, especially when dealing with non-linear correlations (Dimiduk et al. 2019; Mueller et al. 2016). Algorithms based on concepts of artificial intelligence (AI) are known for determining non-linear correlations and have been applied for a range of applications including but not limited to image processing, materials modeling, finance, drug discovery etc. (Chakraborti 2004; Mueller et al. 2016; Ramprasad et al. 2017).

AI algorithms can be used for developing mathematical models and for multi-objective optimization of desired properties. Machine learning (ML) is a component of AI-based approach and has been effectively used for developing mathematical models. Machine learning can be further categorized into supervised and unsupervised machine learning. Supervised ML approaches like artificial neural networks (ANNs), are used for developing predictive models, where a bunch of inputs can predict the output (Muller et al. 2016). Supervised ML algorithms need a training set on which the model is trained, and a testing set on which the model is tested (Muller 2016). Unsupervised ML techniques are used for determining patterns/clusters within a data set and utilize the entire data set; that is, a user does not need to specify input, output, or training and testing set (Muller et al. 2016), whereas multi-objective optimization is performed by evolutionary or genetic algorithms (EA/GA). These topics will be discussed throughout the course of this book (Chakraborti 2004).

1.2 MULTISCALE-MATERIALS MODELING

Advances in multiscale-materials modeling can be subdivided into three categories (Arróyave and McDowell 2019, Ghosh and Dimiduk 2011):

- Historical: Serial paradigm
- Current: Integrated computational materials engineering (ICME) (Horstemeyer 2012)
- Future: Virtual materials design (VMD) (Virtual Materials Design 2021)

During phase 1, researchers focused on developing computational tools to establish a relationship between microstructure and the desired properties of the alloy (Ghosh and Dimiduk 2011). This led to the development of the "CALculation of PHAse Diagram" (CALPHAD) approach in the 1960s (Bale et al. 2016). The CALPHAD approach helped in establishing computational materials science (CMS) in the 1970s–1980s as CMS started getting recognition as a separate discipline of its own (Ghosh and Dimiduk 2011).

Phases 2 (ICME) and 3 (VMD) are aimed at forming a global consortium of experts, who can help develop computational tools that can be helpful in reducing time between discovery and deployment of new materials (Horstemeyer 2012,

Virtual Materials Design 2021). It requires integrating information from literature and experiments and simulations to build a knowledgebase or library like data from microstructure, properties, numerical codes, experimental methods, etc. (Virtual Materials Design 2021). Application of artificial intelligence–based algorithms and statistical toolboxes can be helpful in drawing vital information from vast amount of data available in different format, and repositories (Horstemeyer 2012, Virtual Materials Design 2021).

1.2.1 Initiatives in the United States

In terms of computational materials science, there are two notable initiatives (Ghosh and Dimiduk 2011, Horstemeyer 2012):

- Accelerated Insertion of Materials program (AIM) (2002–2007)
- Materials Genome Initiative (MGI)/Integrated Computational Materials Engineering (ICME)

In this era of rapid economic development, there is an ever-increasing demand for new materials. Rapid industrializations also have led to deterioration of the environment and have invited stringent norms which calls for rapid update on traditional manufacturing processes, search for energy-efficient processes, check on emissions from vehicles, etc. One of the major challenges is to improve upon vehicle designs for a targeted fuel efficiency of 54.5 mpg or more (EPA 2012). As a result, a significant amount of research has been performed in recent years to design magnets that can be used in wind turbines, or in hybrid cars that can help reduce emissions. Another challenge is the supply of critical elements that depends on the geopolitical aspects like rare-earth crisis in 2010, that resulted in a tenfold increase in the price of certain rare-earth elements that are essential for these magnets. At last, the time required between discovery and deployment of a material is between 10 and 15 years. In order to address this, in 2011 U.S. Congress launched the Materials Genome Initiative (MGI)/Integrated Computational Materials Engineering (ICME) initiatives. MGI/ICME will be a platform where experimentalists from major laboratories and researchers from academia and professionals from industries can come together and exchange their knowledge to accelerate the alloy development process (Kalil and Wadia 2011, Markowitz 2009).

The Accelerated Insertion of Materials program (AIM) was funded by U.S. Defense Advanced Research Projects Agency (DARPA) and the U.S. Air Force Research Laboratory (AFRL). The AIM era was from 2002 to 2007, though GE Aviation too conducted research under the AIM program between 2001 and 2004. At present, MGI/ICME approach aims to build upon the AIM program and has been funded by various agencies of the U.S. government since the U.S. Congress launched it in 2011 (Horstemeyer 2012, Markowitz 2009).

One of the key aspects of MGI/ICME initiative was to motivate the researchers to extensively use computational tools, develop prediction models and perform virtual experiments before performing experiments. This has resulted in significant savings in terms of time required for designing a given alloy system (Horstemeyer 2012, Kalil and Wadia 2011).

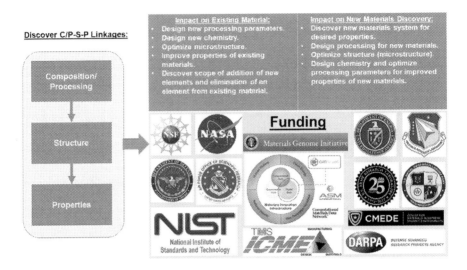

FIGURE 1.1 C/P-S-P linkages: Research areas, impact and potential funding agencies in the USA.

Figure 1.1 shows the potential research area, its impact on existing materials and discovery of new materials, as well as the potential agencies that are funding research in this research domain in the USA.

1.2.2 THE INTEGRATED COMPUTATIONAL MATERIALS ENGINEERING EXPERT GROUP (ICMEG)

ICMEg is an initiative by researchers primarily based in the European Union (Schmitz and Prahl 2014). Their intention is to aid the ICME initiative by forming an expert group. ICME approach is an ongoing process, and any initiative can build upon by setting up a governing body who can set standards for the community. The term *ICMEg* was formed on the lines of "joint photographers expert group" (.jpg/.jpeg) and the "movie pictures expert group" (.mpg/.mpeg). Aims and scopes of ICMEg are similar to that of ICME. ICME was set up in the USA, and ICMEg was formed in Europe to form a group of experts who can set up standards for researchers globally who are working on topics covered under the domain of ICME.

1.3 VIRTUAL MATERIALS DESIGN (VMD)

Determining Composition/Processing-Structure-Properties (C/PSP) linkages is a challenging task, and also the most important step for a researcher who plans to design a new material for a desired/targeted property (or properties) (Ghosh and Dimiduk 2011, Virtual Materials Design 2021). VMD deals with automating the process of materials design by minimizing the scope of human error. VMD aspires on designing a set of computational tools, that can acquire knowledge from the existing information available in the literature, analyze that information, and develop predictive models that can help in simulating experiments in a virtual environment, thus minimizing the actual number of experiments needed to be performed in order to design a new material. Virtual Materials Design is projected as the future of materials design, where computational tools can address challenges at multiple length and time scales involved in materials design. One of the proposed approaches is to utilize machine learning–based surrogate models instead of traditional numerical methods/differential equations. Another challenge is to develop models that can be utilized from HPC to a mobile device (Virtual Materials Design 2021).

1.4 CHALLENGES (MGI/ICME/VMD)

From the research point of view, one has to continue exploring new chemistries, and design new manufacturing protocols. This is to improve upon multiple properties of an existing material, discover new material, improve efficiency of a manufacturing process, or suggest new manufacturing protocols.

From an industrial point of view, most of the existing materials and processes in practice are capable of delivering what it was designed for. That is, these materials and processes are addressing the need of the hour. Most of these materials and manufacturing protocols are time tested. Development of one of the materials or processes may take about 17–20 years. Industrial processes like blast furnace iron making and basic oxygen furnace (BOF) or Linz-Donawitz (LD) steelmaking processes were developed several decades ago. Blast furnace and LD furnace is still an integral part of most of the integrated steel plants around the world. To develop a new iron/steelmaking process will need a consortium of iron and steel industries from around the world. With respect to the development of the FINEX process, Siemens VAI (now Primetals Technologies) and POSCO took decades to design and install a production plant.

Thus, the most important challenge is to deal with the curiosity of the operator who has been manufacturing a certain product by following a certain route for years. Improvement is always desired, but the operator will never let anyone try something new that can halt the entire process that was running efficiently. For example, lowering of silicon content in hot metal (molten iron) from a blast furnace is always desired, but extremely low silicon will result in lowering the temperature of the hearth of the blast furnace, which will affect the fluidity of the hot metal. Similarly, in alloy design, for a multi-component system, there exists several phase equilibria that are extremely sensitive to composition and temperature fluctuations like spinodal decomposition in AlNiCo-based hard magnets.

1.4.1 POTENTIAL SOLUTION THROUGH DATA SCIENCE/ARTIFICIAL INTELLIGENCE

It is extremely important for a data scientist to understand various questions and concerns of an experimentalist or an operator. Then, they must develop their AI/ML algorithms-based predictive models that can address these concerns raised by an experimentalist. Additionally, these models must be robust enough to accommodate additional parameters or concerns raised by the operator. A data scientist must be able to explain various results from their approach in simple statistical terms, plots, and graphs. Most importantly, they must demonstrate the scope of improvement (targeted property, or efficiency of the process) by suggesting a set of parameters, which is within the range of the parameters that an operator is comfortable with. This will help the data scientist in developing/improving their predictive models through inputs from an experienced operator/experimentalist. Once the operator is comfortable with the new parameters, they can implement new suggestions from a data scientist, and also provide valuable feedbacks.

Thus, the goal is to address the concerns raised by a curious experimentalist (operator) through AI/ML-based models and statistics and help them to be confident enough to try new compositions or manufacturing protocol. This can be accomplished by providing them case studies from actual experiments, manufacturing protocols, industrial data, so that they can trust the suggestions from a data scientist.

1.5 THIS BOOK AND ITS CONTRIBUTION TOWARDS MGI/ICME AND VIRTUAL MATERIALS DESIGN (VMD)

This book is primarily based on the application of several concepts of artificial intelligence (AI) and statistics on designing new materials and metallurgical processes. Several concepts of artificial intelligence have been discussed in detail in this book along with the case studies. The case studies include application of AI/ML algorithms on data generated from atomistic simulations (VASP and Schrodinger Materials Design Suite), phase field models, CALPHAD approach, data from experiments (nano-indentation, mechanical, and magnetic properties), and industrial furnaces (blast furnace, LD furnace, rolling mill). Regarding computational cost, all the case studies have been performed on a laptop/desktop computer. In one case study, access to high performance computing (HPC) was needed for atomistic calculations for a portion of the problem. We have demonstrated ways of utilizing mobile devices for executing models developed on a computer. Several open-source softwares have been used in this book. Additionally, we have provided vital information on various AI/ML algorithms, ways of choosing a particular algorithm, and platforms where these algorithms can be used with and without coding. In short, we have covered several aspects of current MGI/ICME in this book along with the proposed future of a materials design "Virtual Materials Design (VMD)" approach.

1.6 AIM AND SCOPE OF CURRENT BOOK

In this section, we describe case studies included in this book.

1.6.1 DISCOVERING AND OPTIMIZING CHEMISTRY/PROCESSING-STRUCTURE-PROPERTY (C/PSP) LINKAGES

1.6.1.1 State of the Art
Determining the relations between chemistry, processing, structure, and property (C/PSP) is a key aspect of materials design. It provides vital insights that will help reduce the time between the discovery and deployment of new materials and improve the properties of existing alloys.

1.6.1.2   Challenge

Although a significant amount of work has been reported on structure-property relationships, processing is still a challenging task. Processing includes optimizing manufacturing protocol, of which heat treatment is most important for alloy design purposes. Processing along with composition variation is even more complex to deal with. In alloy design, any alloy can constitute about 10–12 elements. There exists several phase equilibria, which can be sensitive towards composition and temperature fluctuations like spinodal decomposition. Thus, it is important to understand C/PSP linkages prior to performing experiments.

This work can be applied in discovering new materials as well as improving properties' existing materials. The potential field of applications includes, but is not limited to batteries, superalloys, steels, magnets, etc. Magnets serve the needs of the electronic and automotive industry. Hard magnets are used in designing electric cars, hybrid cars, aerospace industry, electric motors, and renewable energy sources like high-energy density magnets for motors in windmills, etc. Soft magnets find application in the electronic industry. Nickel-based superalloys and aluminum and Tttanium alloys are used for aerospace applications. Titanium alloys are also used for biomedical applications.

1.6.1.3   Research Plan

We have efficiently used AI-based algorithms based on data generated from experiments and information from the literature and simulations performed under the framework of the CALPHAD approach (Jha et al. 2014–22). In this book, we have presented case studies on the following:

1. Soft magnetic FINEMET-type alloys
2. Hard magnetic AlNiCo alloys
3. Nickel-based superalloys
4. Titanium-based alloys for aerospace (high-temperature applications)
5. Titanium-based alloys for biomedical application
6. Aluminum-based alloys

1.6.2   DATA SCIENCE/ARTIFICIAL INTELLIGENCE TECHNIQUES FOR ACCELERATING THE DISCOVERY AND DEPLOYMENT OF NEW MATERIALS

1.6.2.1   State of the Art

Both experiments and simulations generate a large volume of data. Researchers around the world are working on new approaches for using this data to accelerate material discovery and deployment.

1.6.2.2   Challenge

One of the major challenges researchers face is finding correlations between design variables (input parameters) and objectives (targeted properties) and correlations among design variables and objectives and to validate their findings through information reported in open literature. Typically, researchers are dealing with multiple goals/objectives that may be conflicting in nature. Furthermore, a limited amount of information is available in literature on such correlations. Therefore, the use of experimental or simulation data on new materials is a challenging task.

1.6.2.3   Research Plan

An experimentalist wants composition and processing parameters while the funding agency demands multiple optimized properties for implementation. Hence, outcomes from CALPHAD, phase field model, experiments (including microscopy), and simulations can be effectively processed through the application of machine learning algorithms by developing models that can be used in the future for new parameters. Once developed, surrogate models are useful for processing a significant amount of data at a fraction of time relative to CALPHAD, the phase field, or conducting experiments. Then, from our own knowledge of the system and predictions of the surrogate model, we can select a few candidates to carry out experiments. This can be performed manually by intuition or experience with working with that system and also computationally through an algorithm, multi-criterion decision making (MCDM) (Miettinen 1999). In MCDM, an expert puts weight on the set of objectives (properties) that are most important. As in many cases, several properties are conflicting but need to be optimized before using the material for a certain application. It will also be useful to combine specific features at different lengths and timescales. The MCDM can also be useful in determining correlations between different scales of modeling. The MCDM can help minimize the propagation of uncertainty across scales; thus, addressing the limitations that exist in multi-scale materials modeling.

1.6.3   DATA FROM PRODUCTION: INDUSTRIAL FURNACES AND ROLLING MILL

1.6.3.1   Iron-Making Blast Furnace

Data from three (3) different blast furnaces from three different integrated steel plant were analyzed in preparing this case study. Parameters and objectives are different for all of these furnaces. A set of AI-based algorithms were utilized for processing data for all these furnaces.

1.6.3.2   Basic Oxygen Furnace (BOF) or LD (Linz Donawitz) Steelmaking Furnace

Data from an operating LD furnace was used for this case study. A graphical user interface (GUI) was developed in MATLAB® programming language. The GUI can be used for determining additions to be made while secondary steel refining in LD furnace. Kinetic simulations were performed within the framework of CALPHAD approach through Thermo-Calc software utilizing the industrial data and outcomes from GUI/APP. Additionally, a section on determining effect of various alloying elements on the crack susceptibility coefficient of steel has been added in this section.

1.6.3.3   Rolling Mill

Data from an operating industrial rolling mill was used in this case study. Two objectives were examined: roll force and torque. Predictive models were developed for both these objectives through supervised and unsupervised machine learning approaches. Two software packages were used, ESTECO modeFRONTIER and

KNIME. From application point of view, three properties are important namely yield strength (YS), ultimate tensile strength (UTS) and elongation percent . In this book, attempts were made to maximize YS, UTS, and elongation percent through application of evolutionary or genetic algorithms (EA/GA) for two classes of steel obtained from a plate mill and from coiling process. Additionally, effect of various operating parameters in a rolling/coiling mill has been presented.

1.6.4 Nano Mechanics and Atomic Force Microscope (AFM) Imaging

We present a computational platform for developing AI-based models capable of predicting a load-displacement curve and AFM image (Jha and Agarwal 2021). This case study utilized experimental data obtained from nano-indentation experiments and AFM imaging performed on two (2) different materials.

1.6.5 Thermo-Mechanical Treatment

This case study utilized experimental data obtained from the Gleeble testing machine. A steel sample was subjected to a set of predefined strain rates (1, 10, 50, 100/s), at temperatures (800, 900, 1000, 1100°C). Final data include strain rate, strain, temperature, and associated stress. AI-based algorithms were used for developing predictive models for flow stress as a function of strain rate, strain, and temperature. Experimental results were compared with simulations performed in JMATPRO software. Finally, supervised machine learning based prediction models were used for developing stress distribution maps.

1.6.6 Phase Field

This case study utilized data generated through phase field simulations for developing predictive models for grain size as a function of phase field model parameters. Thereafter, we used the concept of parallel coordinate chart for determining the phase field model parameters from the phase field code that will be helpful in achieving these grain sizes.

1.6.7 Atomistic Simulations

Two kinds of software were used for two different case studies.

1.6.7.1 Vienna Atomistic Simulation Package (VASP)

It was used for determining the magnetic state of various phases belonging to the Fe-Si system.

1.6.7.2 Schrodinger Materials Design Suite

MS Jaguar software was used for analyzing infra-red absorption spectra of refrigerants, which is an important parameter for determining the global warming potential (GWP) of a refrigerant (Bochevarov et al. 2013). The initial batch consists of 295 refrigerants, which are used to develop surrogate models followed by multi-objective optimization. A total of 43 new candidates was obtained after design and optimization. Several other candidates were obtained, but their SMILE notation was invalid. Calculations were performed for initial 295 and optimized 43 candidates. It can be observed that the intensity of spectra significantly decreases for 43 optimized candidates when compared with the initial 295 candidates.

Schrodinger Materials Design Suite has been used to discover drugs recently (Schrodinger drug discovery). In the case of the development of a new drug, the analysis of existing drugs on a similar disease may be considered as a starting point. But this analysis involves several parameters, experiments, testing of drugs on test subjects, analyzing effects of drugs on different test subjects and then improving upon the drugs prior to proposing the drugs as an effective solution. In addition, the drug development team should take into account the emergence of new variations and mutations in the future. Therefore, a case study of the application of this software to demonstrate the scope of the new refrigerant design will be important.

1.6.8   EFFICIENT WAY OF CHOOSING A SUPERVISED MACHINE LEARNING ALGORITHM

1.6.8.1   State of the Art

A user relies on open literature, and blogs prior to choosing a supervised machine learning algorithm for developing predictive models. Certain individuals or organizations rely on a particular algorithm or a proprietary software, in-house developed software, or a particular programming language.

1.6.8.2   Challenges

There exist several supervised machine learning algorithms, software, or in-house developed code. This makes choosing a particular approach a complicated task. A popular approach may be effective for a particular problem, but can be misleading for another problem. There exist several start-ups that offer AI-based service, but they usually work on a particular algorithm. Sometimes, a particular algorithm may not be suitable for a particular size of data and number of parameters. Another major problem with any AI-based algorithm is that its prediction accuracy suffers outside the variable bounds.

1.6.8.3   Research Plan

We analyzed performance metrics of a set of AI-based algorithms included in modeFRONTIER software. For test cases, we chose Schittkowski test cases, Case 1–24. These cases are for two-dimensional problems.

Additionally, we have provided a workflow in KNIME software, which is free (open source). A user can use their own data for developing models without coding a single line of code. The test case used here is Schittkowski test problem 47, which has five variables.

This way a user can analyze the performance of various popular approaches. This will help them in choosing a particular algorithm when working on their own problem. A user must note that model parameters were not optimized. An experienced data scientist can improve performance of all of these models by tuning the model parameters.

1.6.9 CALPHAD APPROACH

In this book, we have utilized data generated from CALPHAD-based simulations from FACTSAGE (Bale et al. 2002), Thermo-Calc (Andersson et al. 2002), JmatPro, and PyCalphad/ESPEI. In this section, we have presented a case study on uncertainty quantification and propagation within the framework of CALPHAD approach.

1.7 CASE STUDY #1: UNCERTAINTY QUANTIFICATION AND PROPAGATION STUDIED UNDER THE FRAMEWORK OF CALPHAD APPROACH USING PYCALPHAD AND ESPEI SOFTWARE

PyCalphad, is an open-source software and ESPEI stands for extensible self-optimizing phase equilibria infrastructure (Bocklund et al. 2019, Otis and Liu 2017, Otis et al. 2021). In this book, we have included a case study for uncertainty using PyCalphad software and ESPEI module. All the codes are hosted on the website of PyCalphad and on Github.

A binary alloy system of chromium and nickel was considered for this case study. All the calculations were performed on JupyterHub (sponsored by IBM), a platform provided by the PyCalphad/ESPEI group during their workshop on June 28–29, 2021.

1.7.1 UNCERTAINTY QUANTIFICATION

For the Cr-Ni system, three phases were considered in this study: FCC_A1, BCC_A2, and LIQUID phase. These are the notations from the database hosted on the PyCalphad website. The Markov Chain Monte Carlo (MCMC) algorithm was used for uncertainty quantification.

Corner plots for FCC_A1, BCC_A2, and LIQUID phases are shown in Figure 1.2. In the corner plots, the estimated Cramer-Rao covariance ellipsoids are superimposed on the plots at one and two standard deviations. Regarding computation, the iterations were fixed at 700, and 1,000 MCMC chains were used for fitting all the parameters in this study.

1.7.2 UNCERTAINTY PROPAGATION

After uncertainty quantification, we move ahead towards studying uncertainty propagation. Figure 1.3 shows the variation of enthalpy of mixing for FCC_A1, BCC_A2, and LIQUID phases at 2000 K. One can observe that, the region of highest uncertainty for BCC_A2 is around X(NI) = 0.8. While for FCC_A1, the region of highest uncertainty is between X (NI) = 0.2 and X (NI) = 0.8. While, for

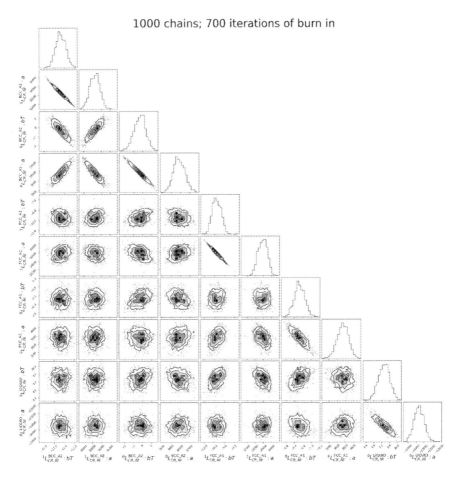

FIGURE 1.2   Corner plots for FCC_A1, BCC_A2, and LIQUID phases.

the LIQUID phase, the region of highest uncertainty is between X (NI) = 0.4 and X (NI) = 0.6. Thermochemical measurements are considered as the foundation for developing accurate thermodynamic models in the CALPHAD community. Thus, uncertainty in the estimation of thermochemical measurements will affect the accuracy of thermodynamic models. From Figure 1.3, one can observe the extent of propagation of uncertainty for "enthalpy of mixing" for Cr-Ni system.

In Figure 1.3, one can observe the region of maximum uncertainty involved with estimation of enthalpy of mixing for BCC_A2, FCC_A1, and LIQUID phases. Thus, we calculated equilibrium calculations for estimating the amount of various phases in the region of maximum uncertainty.

Figure 1.4(a) shows the variation of amount of BCC_A2, FCC_A1, and LIQUID phases for X(NI) = 0.5. The enlarged view of Figure 1.4(a) is shown in Figure 1.4(b). In Figure 1.4(b), one can observe that the thickness of all the lines on the phase diagram varies for different temperatures. In Figure 1.3, we observed that in the region around X(NI) = 0.5, all three phases are uncertain, while maximum

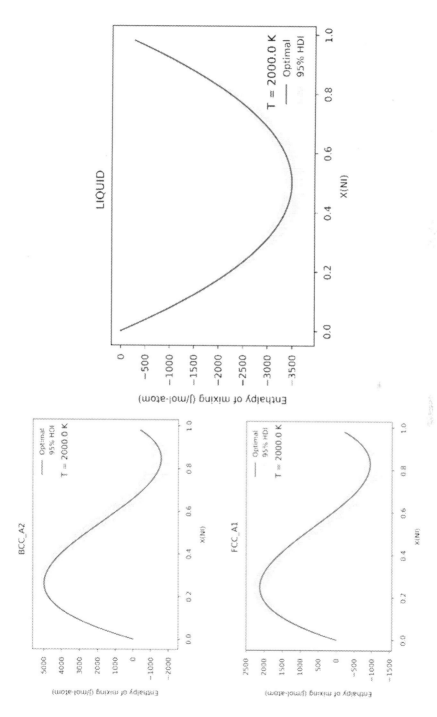

FIGURE 1.3 Enthalpy of mixing for FCC_A1, BCC_A2, and LIQUID phases at 2000 K.

(a)

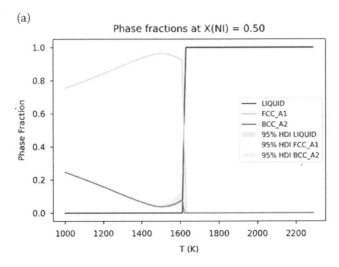

(b)

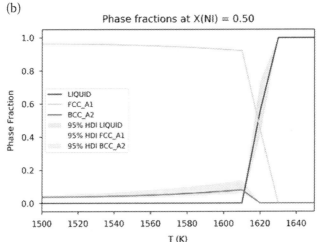

FIGURE 1.4 (a) Phase fraction as a function of temperature for X(NI) = 0.5. (b) Enlarged view showing uncertainty in amount of various phases.

uncertainty is demonstrated for LIQUID phase. In Figure 1.4(b), one can observe that at around 1600 K, uncertainty in the estimation of all three phases is amplified when compared with the other temperature regimes on the figure.

In Figure 1.3, one can observe that in the region around X(NI) = 0.8, all three phases are uncertain, while maximum uncertainty is demonstrated for BCC_A2 and FCC_A1 phase. Figure 1.5(a) shows the variation of various phases as a function of temperature for X(NI) = 0.8, while Figure 1.5(b) shows its enlarged view. In Figure 1.5(b), one can observe the difference in thickness of lines representing various phases at temperature around 1680–1700 K. For other temperatures, line thickness is consistent.

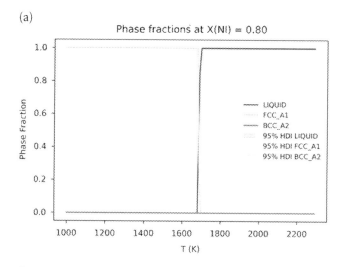

(a)

Phase fractions at X(NI) = 0.80

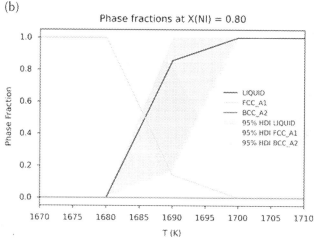

(b)

Phase fractions at X(NI) = 0.80

FIGURE 1.5   (a) Phase fraction as a function of temperature for X(NI) = 0.8. (b) Enlarged view showing uncertainty in amount of various phases.

1.7.3   CONCLUSION: UNCERTAINTY QUANTIFICATION AND PROPAGATION

Regarding uncertainty, Figure 1.2 shows the corner plots for FCC_A1, BCC_A2, and LIQUID phases. In the corner plot, one can visualize uncertainty associated with various thermochemical measurements in the Cr-Ni database for FCC_A1, BCC_A2, and LIQUID phases. These thermochemical measurements constitute the foundation on which various thermodynamic models are based within the framework of CALPHAD approach. Thus, any uncertainty in these measurements will affect the models that are developed by utilizing data from these measurements, like enthalpy of mixing models.

In Figure 1.3, one can observe the effect of propagation of uncertainty from thermochemical measurements to the enthalpy of mixing calculations. One can estimate the region on maximum uncertainty and study uncertainty propagation from enthalpy of mixing model to calculations where enthalpy of mixing calculations is used as an input, like estimation of the phase diagram.

We performed equilibrium calculations for estimating the amount of various phases as a function of temperature for two compositions, $X(NI) = 0.5$ and $X(NI) = 0.8$, the regions identified from Figure 1.3. Figures 1.4 and 1.5 shows the variation of FCC_A1, BCC_A2, and LIQUID phase at $X(NI) = 0.5$ and $X(NI) = 0.8$, respectively. One can observe that in certain temperature regions in both the figures, thickness of line representing the various phases differs a lot, when compared with the thickness of lines throughout the diagram. This demonstrates how uncertainty in estimation of thermochemical measurements can affect the estimation of various phases for a given system. This uncertainty in variation of all the phases are in different temperature regimes for different compositions.

Thus, in this case study, we demonstrated uncertainty propagation within the framework of CALPHAD approach. In physical metallurgy/heat-treatment, there exists several phase equilibria for multicomponent systems. Most of the transformations are diffusion based, and any fluctuation in temperature and composition can affect the amount (volume fraction) of critical phases in the final component. Amount (volume fraction) of a critical phase, and its grain size affect multiple macroscopic properties of an alloy. Thus, uncertainty quantification is extremely important as in materials design there is scope of uncertainty in every stage of manufacture. Uncertainty in measurement of raw material, temperature of furnace, machine readings while performing tests for determining bulk properties, etc. An experimentalist must have proper understanding of various sources of uncertainty. While a data scientist must develop models that are capable of quantifying uncertainty, determine propagation of uncertainty, and most important, find ways of minimizing the effect of uncertainty in the bulk properties of the final product.

In this book, we have presented several case studies where we developed surrogate models using several approaches within the framework of artificial intelligence (AI). Additionally, we carried out several statistical analyses. We have used actual experimental data, data from industries, and simulations to demonstrate ways of minimizing uncertainty propagation.

2 Metallurgical/Materials Concepts

2.1 MATERIALS/ALLOY DESIGN

A researcher must have a problem in hand. Thereafter, they can proceed towards collecting relevant data and analyzing it through statistical tools or through artificial intelligence (AI)–based algorithms. In this chapter, we have stated the underlying physics of various materials system and a few metallurgical processes. This chapter will be helpful in following the case studies reported in this book.

2.1.1 EFFECT OF CHEMICAL COMPOSITION

In Chapter 1, we presented a case study on "uncertainty quantification and propagation" within the framework of CALPHAD approach. For a binary system, we observed the uncertainty propagation with respect to the estimation of "enthalpy of mixing" of BCC_A2, FCC_A1, and LIQUID phases. We also observed that the region of maximum uncertainty was for a given composition of Ni in the Cr-Ni system. Thus, in a binary system, the effect of a slight variation in composition results can be understood with respect to the calculation of enthalpy of mixing of various phases as well as an estimation of various phases as a function of temperature.

In a multi-component system, there exist several equilibria, invariant reactions like eutectic, peritectic, spinodal decomposition, etc., that are extremely sensitive to temperature and composition fluctuation (De et al. 2007, Deva et al. 2008, Deva et al. 2011, Deva and Jha 2016). In structural materials like steel, minor fluctuation in composition of elements like boron, nitrogen, and silicon can have a pronounced effect (Deva et al. 2008, Deva et al. 2011). Any uncertainty in the calculation of thermochemical parameters via experiments or atomistic simulations will propagate when these thermochemical measurements are used within the framework of CALPHAD approach (Otis and Liu 2017, Otis et al. 2021). Figure 2.1 shows the flowchart describing our approach for modeling materials at various scales.

Significant amount of data is generated at each scale of materials design. AI/ML-based algorithm will act as a bridge between various scales and thus help in the discovery and optimization of chemistry-processing-structure-properties linkages in existing materials as well as discover new materials. This will be helpful in improving properties of existing materials, accelerated discovery of new materials, and their deployment into service.

2.1.2 EFFECT OF PROCESSING

Processing or manufacturing protocol can include heat treatment, mechanical working, magnetization, etc. Temperature greatly influences the stability of various

DOI: 10.1201/9781003167372-2

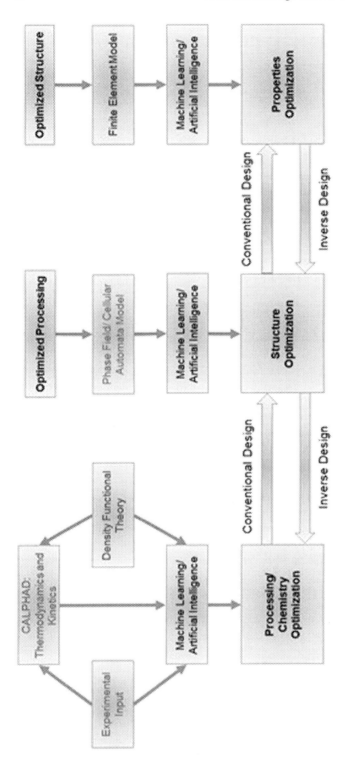

FIGURE 2.1 Optimizing Processing (Chemistry)-Processing-Properties for designing materials.

phases desired for achieving targeted properties. Heat-treatment schedule design is critical in stabilizing a certain phase while suppressing another phase, which is considered detrimental for a particular alloy application (De et al. 2011, 2007). Throughout this book, we have included case studies on stabilizing and suppressing various stable and metastable phases within the scope of CALPHAD approach. Typically heat-treatment schedule includes solutionizing, annealing: isothermal and non-isothermal, quenching, and tempering. Solutionizing involves reheating a sample above a temperature where all the precipitates dissolve. Isothermal annealing holding involves exposing a sample to certain prescribed temperature, mainly for nucleation and growth of a certain desired phase. Non-isothermal annealing involves heating or cooling a specimen at a prescribed rate for nucleation and growth of a certain desired phase. Quenching involves fast cooling of a sample, in order to suppress formation of any particular undesired phase. Tempering is performed at elevated temperature for stress relaxation (De et al. 2007).

Mechanical working like cold rolling, hot rolling, etc. are routinely performed. Mechanical working above recrystallization temperature is known as hot working, and below recrystallization temperature is known as cold working. Recrystallization temperature is about 0.6 Tm, where Tm is the melting temperature (De et al. 2011). Mechanical working is performed for introducing dislocations, which acts as nucleation sites for precipitation of various stable and metastable phases during aging, particularly in aluminum alloys.

Magnetization: We have included case studies for both hard and soft magnets. Hard magnets like AlNiCo alloys are magnetized during heat treatment stages. Magnetization is an important step in magnetic alloys as magnetic properties are dependent of alignment of dipoles in a particular direction (Dilon 2014).

2.1.3 EFFECT OF STRUCTURE

The properties of polycrystalline materials are often influenced by grain size (De et al. 2007). From mechanical properties such as yield strength; hardness; breaking strength; creep resistance to electrical, optical, and magnetic properties; and grain size has a direct influence on it. Grain size reduction is commonly used to improve the yield strength of metals, as the smaller grains create more obstacles to dislocation motions and thus improve strength. The particle size-strength relation follows the Hall-Petch equation. On the other hand, to design a creep-resistant material, a larger grain-size system will benefit. At a higher temperature, grain boundaries become the weakest link and promote slip deformation. Small grain size material is also more susceptible to corrosion. Grain boundaries loosen atomic structures due to high interfacial energies and become the preferred sites for the onset of corrosion. As a result, the preferred grain size depends entirely on the application. It is therefore imperative that the grain size is controlled or optimized especially when designing a material for any specific application (De et al. 2007, 2011).

2.1.4 Effect of Composition, Processing, and Structure on Properties (C/P-S-P)

Processing-structure-property (PSP) linkages are being worked upon by several research groups around the world. In this book, we have added another important component of materials design, "Composition"). Thus, we have framed this term composition-processing-structure-properties (C/P-S-P). We have presented several case studies in this book using experimental data and data from simulations for explaining the C/P-S-P linkages.

2.2 CALPHAD APPROACH

In this book, several case studies utilize information extracted from CALPHAD-based simulations. In this section, we have described the Kampmann-Wagner Numerical (KWN) approach. It is followed by a short description on various CALPHAD-based software.

2.2.1 Precipitation Kinetics: Nucleation and Growth of Grains/Crystals of Desired Phase

The Kampmann-Wagner Numerical (KWN) approach (Cinkilic et al. 2020, Kampmann et al. 2000, Langer and Schwartz 1980, Sarafoglou et al. 2019, Wagner and Kampmann 1991, Zhang et al. 2019): The precipitation module TC-PRISMA in Thermocalc is based on Langer-Schwartz theory. It is based on the Kampmann-Wagner Numerical (KWN) approach for simulating precipitation kinetics of a precipitate and for prediction of particle size distribution (PSD) over the full course of precipitation for multicomponent systems and multiphase alloys (Andersson et al. 2002).

In Chapter 8, the matrix phase is FCC_L12 and the precipitate is AL3X. AL3X is Thermocalc notation for Al_3Sc phase. The molar volume of precipitation originates from the Thermocalc database. The grain aspect ratio was fixed at 1.0, whereas the mobility enhancement factor was fixed at 5.

Initial number of nucleation sites (N_0) were calculated from dislocation density, which was fixed at $6.0 \times 10^{12}/m^2$. Interfacial energy was set at 0.135, 0.130, 0.125, and 0.115 J/m^2 for 300, 350, 400, and 450°C, respectively.

Time-dependent nucleation rate (J_t) (Russell and Yamauchi 1980) is expressed in equation 2.1.

$$J_t = J_s \exp\left(\frac{-\tau}{t}\right) \tag{2.1}$$

In equation 2.1, τ is incubation time. This can be further simplified as shown in equation 2.2, t is the time and J_s is the steady-state nucleation rate, as shown in equation 2.3 theory (Agren et al. 2015, Bardel et al. 2014, Bonvalet et al. 2015, Russell and Yamauchi 1980).

$$\tau = \frac{1}{\theta Z^2 \beta^*} \qquad (2.2)$$

$$J_s = Z\beta^* N_0 \exp\left(\frac{-\Delta G^*}{kT}\right) \qquad (2.3)$$

In equations 2.2 and 2.3, Z stands for Zeldovich factor, β^* is the attachment rate of solute atoms to the precipitate (AL3X), N_0 denotes number of nucleation sites available in the beginning, ΔG^* is the Gibbs energy of formation of precipitate, k is Boltzmann's constant, and T is absolute temperature (Andersson et al. 2002, Bardel et al. 2014, Russell and Yamauchi 1980). In equation 2.2, θ can vary but in Thermocalc it is fixed at 2 (Andersson et al. 2002).

The number density of precipitates in the beginning can be shown in equation 2.4, while number density (N_t) at any instant of time (t) can be expressed in equation 2.5 (Andersson et al. 2002).

$$N_0 \exp\left(\frac{-\Delta G^*}{kT}\right) \qquad (2.4)$$

$$N_t = \int J_t dt \qquad (2.5)$$

ΔG^* is the Gibbs energy of formation of a precipitate can be expressed as in equation 2.6. $\Delta G_m^{FCC-Al_3Sc}$ is the molar Gibbs energy of formation of Al_3Sc nano-crystals from the FCC matrix, $V_m^{Al_3Sc}$ is the molar volume of Al_3Sc nanocrystals (Cinkilic et al. 2020), while σ is the interfacial energy between the FCC matrix phase and precipitate Al_3Sc phase.

$$\Delta G^* = \frac{16\pi}{3} \frac{\sigma_{int}^3}{\left(\frac{\Delta G_m^{FCC-Al_3Sc}}{V_m^{Al_3Sc}}\right)^2} \qquad (2.6)$$

Zeldovich factor (Z), and β^* can be expressed as shown by equations 2.7 and 2.8, respectively (Agren et al. 2015, Andersson et al. 2002, Bonvalet et al. 2015, Chen et al. 2008).

$$Z = \frac{V_m^{Al_3Sc}}{2\pi N_A R^{*2}} \sqrt{\frac{\sigma}{kT}} \qquad (2.7)$$

$$\beta^* = \frac{2\pi R^{*2}}{(l^{Al_3Sc})^4} \left[\sum \frac{\left(\overline{X_i^{Al_3Sc}} - \overline{X_i^{FCC}} \right)^2}{\overline{X_i^{FCC}} D_i^{FCC}} \right]^{-1} \tag{2.8}$$

In equation 2.8, $\overline{X_i^{FCC}}$ and $\overline{X_i^{Al_3Sc}}$ are the equilibrium composition of element i in the FCC and in the Al$_3$Sc phase, respectively, while D_i^{FCC} is the chemical diffusion coefficient of element i in the FCC phase (Agren 2015, Bonvalet et al. 2015, Chen et al. 2008, Li et al. 2017).

For Al$_3$Sc crystals, critical radius ($R*$) (Cinkilic et al. 2020, Kampmann et al. 2000, Sarafoglou et al. 2019, Wagner and Kampmann 1991) and time-dependent radius ($R_t^{Al_3Sc}$) (Chen et al. 2008, Li et al. 2017) can be expressed as in equations 2.9 and 2.10, respectively.

$$R^* = -\frac{2\sigma V_m^{Al_3Sc}}{\Delta G_m^{FCC-Al_3Sc}} \tag{2.9}$$

$$\frac{dR_t^{Al_3Sc}}{dt} = \frac{D_i^{FCC}}{\xi_{i,t} R_t^{Al_3Sc}} \frac{X_{i,t}^{FCC} - \overline{X_i^{FCC}}}{\overline{X_i^{Al_3Sc}} - \overline{X_i^{FCC}}} \tag{2.10}$$

In equation 2.10, $X_{i,t}^{FCC}$ is the composition of supersaturated FCC matrix phase and $\overline{X_i^{Al_3Sc}}$ is the equilibrium composition of Al$_3$Sc phase. Growth of Al$_3$Sc crystals is dependent on $X_{i,t}^{FCC}$. Mass balance equation can be expressed as shown in equation 2.11 (Bonvalet et al. 2015).

$$X_{i,t}^{FCC} = \left(X_{i,0}^{FCC} - V_f^{Al_3Sc} \overline{X_i^{Al_3Sc}} \right) / \left(1 - V_f^{Al_3Sc} \right) \tag{2.11}$$

In equation 2.10, parameter $\xi_{i,t}$ can be expressed as in equation 2.12, while parameter $\lambda_{i,t}$ in equation 2.12 can be calculated by solving equation 2.13 (Bonvalet et al. 2015, Chen et al. 2008, Li et al. 2017).

$$\xi_{i,t} = 1 - \lambda_{i,t}\pi \exp(\lambda_{i,t}^2) \text{erfc}(\lambda_{i,t}) = \frac{1}{2\lambda_{i,t}^2} \frac{X_{i,t}^{FCC} - \overline{X_i^{FCC}}}{\overline{X_i^{Al_3Sc}} - \overline{X_i^{FCC}}} \tag{2.12}$$

$$2\lambda_{i,t}^2 - 2\lambda_{i,t}^3 \sqrt{\pi} \exp(\lambda_{i,t}^2) \text{erfc}(\lambda_{i,t}) = \frac{X_{i,t}^{FCC} - \overline{X_i^{FCC}}}{\overline{X_i^{Al_3Sc}} - \overline{X_i^{FCC}}} \tag{2.13}$$

Coarsening rate of precipitate can be predicted from the growth equation shown in equation 2.14 (Li et al. 2017, Perez et al. 2008, Rougier et al. 2013).

$$\frac{dR_t^{Al_3Sc}}{dt} = \frac{8}{27}\frac{\sigma V_{at}^{Al_3Sc}}{(R^{Al_3Sc})^2 kT}\frac{D_i^{FCC}\overline{X_i^{FCC}}}{X_i^{Al_3Sc} - X_i^{FCC}} \tag{2.14}$$

In equation 2.14, $V_{at}^{Al_3Sc}$ is the mean atomic volume of Al_3Sc phase, while N_A is the Avogadro's number.

$$V_{at}^{Al_3Sc} = \frac{V_m^{Al_3Sc}}{N_A} \tag{2.15}$$

Time dependent mean radius and volume fraction of Al_3Sc nanocrystals at any time step can be calculated through equations 2.16 and 2.17, respectively (Agren et al. 2015, Bonvalet et al. 2015, Li et al. 2017, Perez et al. 2008, Rougier et al. 2013).

$$\overline{R_t^{Al_3Sc}} = \frac{\Sigma N_t R_t^{Al_3Sc}}{\Sigma N_t} \tag{2.16}$$

$$V_{f.t}^{Al_3Sc} = \Sigma N_t \frac{4\pi (R_t^{Al_3Sc})^3}{3} \tag{2.17}$$

2.2.2 CALPHAD-BASED SOFTWARE AND DATABASE: APPLICATION AND LIMITATIONS

2.2.2.1 THERMOCALC (Andersson et al. 2002)

We have used Thermocalc software for a number of case studies. The workings of Thermocalc software will be described in these case studies. We have used equilibrium module, TC-PRISMA module, and Property model calculator.

2.2.2.2 FACTSAGE (Bale et al. 2002)

We have presented results by performing calculations using the equilibrium module and reaction module. We also used the fact optimal module for demonstrating use of FACTSAGE for optimization.

For exploring new materials, there is more room in the equilibrium module to develop new alloys, particularly for multicomponent system as the output result satisfies the criterion for free energy minimization. It calculates the conditions for multiphase, multi-component equilibria, with a wide variety of tabular and graphical output modes, under a large range of constraints and accesses both compound and solution databases. Gases can be assumed to be real or ideal as per the user requirement. There is a provision for expansion and compression for various phases.

The reaction module also provides the most stable product at any temperature, but the procedure is more ideal as in it all the gases are considered ideal and it also ignores expansion and compression of solids and liquids.

2.2.2.2.1 Pareto-Optimal Set in CALPHAD-Based Software FACTSAGE

This is a small case study using CALPHAD-based software FACTSAGE. In Chapter 3, we have discussed concepts of Pareto-optimality under multi-objective optimization. FACTSAGE software can be used to obtain Pareto set (solutions). In this section, we have demonstrated this feature of FACTSAGE through a case study on "Reduction of $Fe_2O_3$with CH_4".

Problem formulation: Maximize: (Fe_liq + H_2_gas) and Minimize: Cost

Cost: Assume cost of Fe_2O_3 is 40\$/Ton and CH_4 is 400\$/Ton.

Physical Constraints to Be Used in FACTSAGE Software

- The upper and lower bounds for Fe_2O_3 and CH_4 were set at 1 and 0.
- $CH_4 > 1.4495*Fe_2O_3$ (so that there is no unreduced iron left and provide relaxation in search space)
- $CH_4 < 3.0* Fe_2O_3$ (To avoid wastage of costly CH4 gas)
- Temperature: **1540–1560°C**
- Number of Quasi Random calculations: 50, Number of Factsage calculations: 25

Pareto-optimal solutions have been tabulated in Table 2.1. In Table 2.1, the value in bold is the best possible solution suggested by FACTSAGE software.

Pareto-optimal solutions have been plotted in Figure 2.2(a). In Table 2.1, the best possible solution was suggested by FACTSAGE. Another optimization problem was performed, where the goal was to maximize both Fe (liquid) and H2 gas. In Figure 2.2(b), one can observe that the Pareto-front is a single point.

In Chapter 3 and the rest of the book, we have presented case studies where we used an evolutionary algorithm (EA) or a genetic algorithm (GA) for performing

TABLE 2.1
Pareto-optimal Solutions: Case Study on FACTSAGE Software

Fe2O3	CH4	Fe+H2	$
0.4	0.6	0.7	87.14
0.38	0.62	0.8	90.70
0.36	0.64	0.9	94.55
0.34	0.66	1.01	98.75
0.32	0.68	1.12	103.33
0.31	**0.69**	**1.17**	**105.79 (BEST)**
0.30	0.70	1.22	108.36
0.28	0.72	1.33	113.91
0.26	0.74	1.44	120.05

(a)   (b)

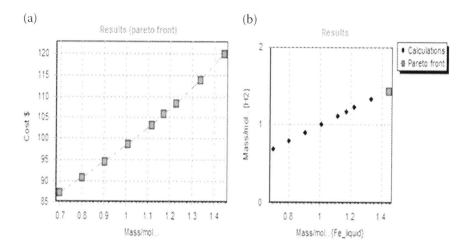

FIGURE 2.2   Pareto-optimal solutions: (a) Maximize (Fe (liquid) + H$_2$) and minimize cost. (b) Maximize Fe (liquid) and maximize H$_2$ gas.

multi-objective optimization. In Figure 2.2(b), Pareto-front is just one point, which happens a lot of times when the algorithm is stuck in a local maxima/minima in the search space. Artificial intelligence–based algorithms, evolutionary algorithm (EA) or genetic algorithm (GA) can be helpful in solving this problem and a user can get a Pareto-front with multiple solutions.

This case study was performed for two reasons:

- To demonstrate that optimization of two objectives can be performed in CALPHAD-based software FACTSAGE.
- Sometimes an algorithm provides just one solution, which is misleading. Multi-objective optimization must provide with a well-defined set of multiple distinct solutions. Thus, a user must change the parameters of the algorithm to check upon improvements. As per our experience, sometimes this doesn't work at all. In this case, a user must choose a different algorithm prior to drawing conclusions.

In multi-objective optimization, a user cannot get just one solution; thus, they must not conclude that they achieved the best possible solution, as the algorithm just gives them one solution.

In other chapters, we have mentioned it through a case study, particularly in hard magnetic AlNiCo type alloys, where several EA/GA algorithms were stuck and provided with just one solution.

2.2.2.3   JMatPro®: (JMatPro 2021)

JMatPro is a commercial software based on the CALPHAD approach. In addition, with phase transformations, it is helpful in determining a number of mechanical properties. Mechanical properties can include room-temperature strength/hardness, high-temperature strength/hardness, flow stress curves, rupture strength, creep and rupture life, processing map, and fracture toughness.

In alloy design and especially new alloy discovery, one must use both JMatPro and Thermocalc. JMatPro can be used alone too. Thermocalc databases include more stable and metastable phases when compared with JMatPRO databases. But, one cannot simulate mechanical properties in ThermoCalc. Thus, new compositions can be generated in Thermocalc and JMATPRO together. Then for new compositions, one can estimate the equilibrium amount of phases, study phase transformations, and simulate mechanical properties in JMatPro software.

We have presented two (2) case studies in Chapters 9 and 15, where we have presented data generated from JMatPro.

2.2.2.4 PyCalphad/ESPEI (Otis and Liu 2017, Otis et al. 2021)

It's an open-source software developed within the framework of CALPHAD approach. ESPEI stands for extensible self-optimizing phase equilibria infrastructure. In this book, we have included a case study for uncertainty quantification and propagation in Chapter 1 using the PyCalphad and ESPEI module.

2.2.2.4.1 Phase Diagram in PyCalphad

PyCalphad is a relatively new software, and in the future it has the potential to be competitive as it is an open-source platform. In a phase diagram, there are some limitations right now, as the developers are still working on adding the invariant lines on the phase diagram. Currently, these lines are missing, as can be seen from the diagram for the Fe-Si system in Figure 2.3. This software can be used right now for learning purposes as all the codes are written in the Python programming language. Code and a few databases are free for users, thus a user can install PyCalphad on their own computer for free.

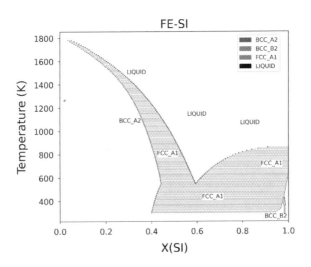

FIGURE 2.3   Fe-Si phase diagram plotted through PyCalphad software.

For uncertainty quantification using ESPEI module, calculations can be computationally intensive. A user may need access to a computer with a decent GPU, or a supercomputer. Additionally, a user needs to develop several files needed as input for the software. PyCalphad/ESPEI software will be great for users who are into programming and users who want to learn Python programming language.

2.3 NANOMECHANICS AND NANOTRIBOLOGY

2.3.1 NANOINDENTATION

Nanoindentation is extensively used to determine the mechanical properties of materials (Oliver and Phar 1992). Atomic force microscopy (AFM) imaging helps in analyzing the surface morphology at the nanoscale level, and in studying material deformation during indentation (Liu et al. 2020). A user requires AFM while checking the indent image to study deformation behavior (Liu et al. 2020, Oliver and Phar 1992). Figure 2.4 provides a schematic view of the nanoindentation experiment.

In a nanoindentation test, a set of load, loading rate, and holding time values is predefined. A load (P) is applied on the indentor as shown in Figure 2.4. The indentor is pressed upon the materials in accordance to the experimental parameters preset by the user. Data generated from the nanoindentation test is presented as a load displacement curve. This curve is used for estimating mechanical properties of a material.

Mechanical properties of interest may include hardness (σ), stress exponent factor (η), strain rate sensitivity (m), creep compliance, etc. These values are calculated as follows.

Stress exponent factor (η): It is a measure of creep resistance. The value of n indicates the creep mechanism. For diffusion creep, n is between 0 and 1, and for dislocation Creep: n = 3 to 8 (Oliver and Phar 1992). Following equations (2.18 to 2.21) were used in determining the values of stress exponent factor (η). Load (P) and contact depth (h_c) are shown in Figure 2.4.

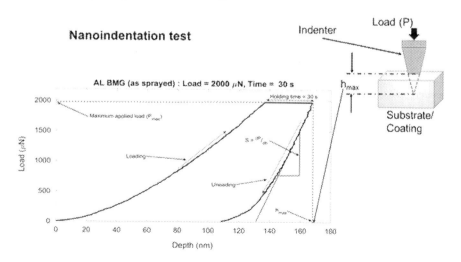

FIGURE 2.4 Schematic representation of nanoindentation test.

$$\dot{\varepsilon} = A\sigma^n \tag{2.18}$$

$$\dot{\varepsilon} = \frac{1}{h}\frac{dh}{dt} \tag{2.19}$$

$$\sigma = \frac{P}{24.5_*h_c^2} \tag{2.20}$$

$$n = \frac{\partial ln\,(\dot{\varepsilon})}{\partial ln\,(\sigma)} \tag{2.21}$$

Strain rate sensitivity (*m*): It is defined as the resistance of a material towards deformation, especially when strained. This is fairly common with aluminum alloys. It can also be viewed as an inverse of the stress exponent factor (equations 2.22 and 2.23).

$$\sigma = C\dot{\varepsilon}^m \tag{2.22}$$

$$m = \frac{\partial ln\,(\sigma)}{\partial ln\,(\dot{\varepsilon})} \tag{2.23}$$

Compliance: Assistance offered by the material to the indentor during indentation (penetration) is referred to as compliance. It is the inverse of stiffness. Therefore, with the increase in load, the resistance offered will decrease and therefore compliance should increase with the increase in load. Creep compliance can be calculated by equation 2.24.

$$J_c(t) = \frac{8\,tan\,(\alpha)\,h^2(t)}{\pi P_o} \tag{2.24}$$

Where,
A is the semi-angle of Berkovich tip, h(t) is the instantaneous indentation depth, P_o is the Peak load.

2.3.2 ATOMIC FORCE MICROSCOPY (AFM) IMAGING

AFM imaging and the generation of topographical and gradient images are illustrated in Figure 2.5. There is an interaction force between the tip of the AFM and the face of the specimen when the tip is near the surface. When performing AFM imaging, this interaction force can be estimated. The AFM tip is mounted on the end

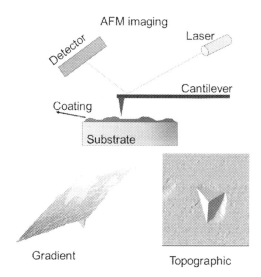

FIGURE 2.5   Schematic representation of AFM imaging.

of a cantilever and the cantilever is moved parallel to the surface to estimate the surface profile. Cantilever moves along the x-y direction. A laser system follows the movement of the cantilever and this laser detection system is able to estimate the interacting force between the tip of the AFM and the surface. When performing AFM imaging, surface roughness affects the interaction force between the tip of the AFM and the surface. For AFM, a sensor set point is defined at the start. A laser detection system compares the measured signal to the sensor's set point. During the scan, if it detects a change, it resets the distance from the AFM tip to the surface in the z direction. The difference between the measured signal and the sensor set point is called an "error signal". This error signal is set to zero by changing the distance between the end of the AFM and the surface. Therefore, AFM imaging registers two types of signals, (1) z correction and (2) error signal. AFM imaging provides two types of surface or image profile, namely "TOPOGRAPHY" and "GRADIENT", where the topography is the z-correction, whereas the gradient is the error signal (SPM Modi 2021, Jha and Agarwal 2021). The AFM configuration assigns the terminology "FORWARD" and "REVERSE" according to the starting point from which the cantilever begins to move.

2.3.3   MODELING AND SIMULATION: CASE STUDY IN THIS BOOK

Chapter 4 includes several case studies on utilizing data from nanoindentation test and AFM imaging for developing AI-based predictive models. These models can be used in the future for simulating the nanoindentation test and AFM imaging for new experimental test conditions.

2.4 MAGNETISM AND MAGNETIC TERMINOLOGIES

In this book, we have presented case studies on both hard magnetic AlNiCo type alloys, and soft magnetic FINEMET type alloys. It is important to go through various concepts related to magnetism for better understanding of the test cases.

Several magnetic terminologies have been used in this book (Buschow and de Boer 2003, Cullity and Graham 2009, Herzer 1993, Périgo et al. 2018, Tsepelev and Starodubtsev 2021).

These terminologies are as follows.

2.4.1 MAGNETIZATION (M) (BUSCHOW AND DE BOER 2003)

The magnetization is an intrinsically magnetic property illustrated by the density of magnetic moments per unit volume. The magnetic moment determines a magnet's interaction strength with a magnetic field. Magnetic poles (p) have traditionally been found in pairs (dipole). In a uniform magnetic field, the dipole moment aligns with the applied field in its lowest energy state. A torque acts on the magnet in alignment with the field. The (pole * length) is called the magnetic moment, since it is the moment of the torque applied to the magnet in a uniform field. Ferromagnets spontaneously align atomic moments when the temperature is lower than the magnetic control temperature (Curie temperature). A schematic representation is shown in Figure 2.6.

2.4.2 MAGNETIC FIELD (H) (BUSCHOW AND DE BOER 2003)

The magnetic field (H) is the vectorial field describing the contributions of electrical currents and magnetic dipoles to the intensity of magnetic interactions in space. Magnetic fields may be generated through superconducting coils or electromagnets. Its output is measured in Ampere/metre or Oersted. The magnetic field strength is measured in Tesla and is also known as magnetic field strength, magnetic field, or H field. The range of significance for our considerations is the range from a few nano Tesla (small coercivity) to about tens of Tesla (superconducting magnets).

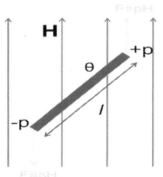

FIGURE 2.6 Magnetization (M).

2.4.3 Magnetic Flux Density (Dilon 2014)

A vectorial field describing the contributions of electric currents and magnetic dipoles to the strength of magnetic interactions in space. It is also known as magnetic induction or B field. Researchers frequently use B and H interchangeably, but B and H differ in free space due to the permeability of free space. Magnetic induction is made up of two components: magnetic field strength and magnetization. As CGS units are not streamlined, it is difficult to convert them into energies when calculating energy products for permanent magnets and core losses for soft magnets. It is necessary to use SI units to make computations using magnetic terms (and convert them to CGS if necessary). 4π in CGS has units (converting M in emu/cc to $4\pi M$ in G).

2.4.4 Core Loss (Pcv) (Herzer 1993)

Core loss is an important characteristic of magnetic materials. From an application standpoint, a soft magnetic material is switched at constant frequency and induction amplitude. Area of the resultant hysteresis loop is equal to the core loss (or energy lost by the core per cycle). The core loss varies depending on the magnitude of the induction applied and the switching frequency.

It has three contributory mechanisms:

- Hysteresis losses: Due to anisotropies in the material. It generally dominates at low switching frequencies and can be controlled by microstructure, composition, etc.
- Classical eddy current losses from induced currents, as a result of variation of the B field. This can be reduced by increasing the resistivity and reducing the thickness of the laminate.
- Dynamic eddy current losses (abnormal losses) are caused by local motion of the magnetic domain walls. It generally predominates at high switching frequency.

Nanocrystalline alloys exhibit some of the smallest core losses on record.
Core losses tend to increase with:

- An increase in applied induction amplitudes. All moments are aligned at high applied induction amplitudes as the material approaches saturation.
- An increase in switching frequency because the contribution of eddy currents to losses depends heavily on the frequency.

2.4.5 Coercivity (Hc) (Cullity and Graham 2009, Buschow and de Boer 2003)

Coercivity is the value of the inverse magnetic field required to bring the net magnetization of a magnetic material to zero or it is the magnetic field required to demagnetize a magnet. Hard or permanent magnets possess large coercivity. This feature is used for power storage and the generated magnetic field can be used for work. Soft magnets must possess a small coercivity, as it is used in applications where switching occurs multiple times per second. Each cycle results in energy loss, a portion of which is proportional to the coercivity. In hard magnets due to larger coercivity, the magnetization does not switch due to external magnetic field.

Coercivity is closely related to the hysteresis losses that predominate at low switching frequencies. Nanocrystalline soft magnetic alloys have improved hysteresis losses compared to coarse-grained materials simply due to their microstructure.

(Fe,Si)-based alloys include silicon steels (few tens of A/m coercivity), some amorphous alloys (few tenths to few A/m coercivity), and Finemet (with few tenths of A/m coercivity). Fe-based alloys include steels (near 100 A/m coercivity), high magnetization amorphous alloys (near 10 A/m coercivity), and Nanoperm (a couple of A/m coercivity). FeCo-based alloys include Hiperco (few hundred A/m coercivity) and HiTperm (few tens of A/m coercivity). In each case, nanocrystalline alloys are associated with a lower coercivity when compared to coercivity compared to their large crystalline alloy counterparts. The reason for the reduced coercivity is due to a change in the dominant magnet switching mechanism. For grains over 100 nm, the switching mechanism is hampered by pinning the magnetic field wall onto the grain boundaries. In the case of grains less than 100 nm, the switching mechanism is described by the random anisotropy model. A benefit of nanocrystalline alloys with the same coercivity as coarse-grained alloys is the increase in resistivity (by a factor of 3). This translates into smaller losses at higher switching frequencies because of the reduced contribution of eddy currents.

2.4.6 MAGNETIC ANISOTROPY (HERZER 1993)

Anisotropy of a material has a strong effect on the shape of the hysteresis loop. Higher values of magnetic anisotropy result in larger remanent magnetization and coercivities. Magnetization gives preference to certain crystallographic directions (easy axes) of a crystalline magnetic material. Additional energy is required for magnetization to align with other crystal directions (hard directions). Thus, magneto-crystalline anisotropy can be defined as the energy needed to move the magnetization from the easy axis to other directions in the crystal (intrinsic property).

Preferred orientation of magnetization (M) due to various anisotropies is as follows:

- Magneto-crystalline (M orientation with respect to crystal axes): The atomic structure of a crystal (depends on composition, crystal structure, atomic ordering)
- Magnetoelastic (M orientation with respect to stress fields): Stress fields combined with the magnetostrictive effect.
- Shape (M orientation with respect to sample shape features and H direction): Magnetostatic forces resulting from magnetic poles on the surface of non-spherical magnetic particles (discontinuity of M at the surface).
- Induced (M orientation with respect to directional ordering in sample): Short-range chemical ordering due to field annealing techniques.

2.4.7   MAGNETOSTRICTION (Λ_s) (BUSCHOW AND DE BOER 2003)

It's an intrinsic magnetic property. That's the stress-dependent part of magneto-crystalline anisotropy. It determines the change in shape of a magnetic material when the magnetization is changed. The magnetostriction of a material can be reduced to a value close to zero, which will translate into a reduction in its contribution to core losses.

2.4.8   PERMEABILITY (M) (CULLITY AND GRAHAM 2009, BUSCHOW AND DE BOER 2003)

The permeability is a measure of response of magnetic flux density of a material to the magnetic field. It is finite in free space ($\mu_0 = 4\ \pi\ 10^{-7}$ kg m A^{-2} s^{-2}). In a magnetic material, the permeability (μ) is often $10^2 - 10^5$ times higher than μ_0.

High permeability means that a small field is necessary for switching the material; thus, high permeability is needed for soft magnets.

2.4.9   CURIE TEMPERATURE (CULLITY AND GRAHAM 2009)

Pierre Curie observed that ferromagnetic substances displayed a critical temperature transition, beyond which the substances lost their ferromagnetic behavior. It was known as the Curie Point. Curie temperature is the magnetic operating temperature. Each magnetic material or phase has a different Curie temperature, where the magnetization drops continuously to the Curie temperature (higher-order phase transformation). Above the Curie temperature, there is sometimes residual magnetization because of the short range order of atomic moments in an applied field.

Nanocrystalline soft magnets are composed of two phases, a crystal phase with high Curie temperature and an amorphous matrix with low Curie temperature.

2.4.10   HARD MAGNETS (JHA ET AL. 2017, MCGUINESS ET AL. 2015)

Hard magnetic AlNiCo alloys are extensively used ever since its discovery due to excellent high temperature and corrosion (McGuiness et al. 2015). AlNiCo magnets have remanence comparable to rare-earth magnets, but coercivity and thus magnetic energy density is lower than rare earth magnets. Any improvement in coercivity and magnetic energy density will be helpful in filling the gap between the rare-earth magnets and current AlNiCo magnets. In Chapter 5, we have presented a case study on AlNiCo-based magnets. Effect of various elements on AlNiCo-based magnets have been tabulated in Table 2.2.

2.4.11   SOFT MAGNETS (BUSCHOW AND DE BOER 2003, HERZER 1993)

Soft magnetic FINEMET alloys are used for high saturation magnetic flux density, low core losses, low magnetostriction, excellent temperature characteristics, small aging effects, and excellent high-frequency characteristics of mobile phones, transformers, noise reduction devices, etc. From an application point of view, a researcher has to deal with multiple properties that are conflicting. An improvement of one property may result in deterioration of other properties. Additionally, processing

TABLE 2.2

Magnetic Alloys: Effect of Various Alloying Elements (Dilon et al. 2014)

Element	Stabilizes	Magnetic Properties Affected	Action Taken
Co	γ phase	It increases H_c and Curie temperature.	Increase solutionization anneal temperature
Ni	γ phase	Increases Hc (less than Cobalt) while decreases Br.	
Al	α phase	Expected to affect Hc positively.	Helps in reducing solutionization anneal temperature.
Cu	α phase	• Increases both Hc and B_r. • In AlNiCo 8 and AlNiCo 9 alloys, Cu precipitates out of the α2 phases into particles and is responsible for the magnetic separation between α1 and α2 phases. • Increase in phase separation leads to an increase in Hc.	
Ti	α phase	• Increases Hc while decreases Br. • Reacts with impurities such as C, S, and N and purifies the magnet by forming precipitates with these elements. • Helps in grain refining. • Detrimental for columnar grain growth. Majority of grains are aligned perpendicular to the chill plate due to columnar grain growth and large shape anisotropy can be achieved if spinodal decomposition occurs in this direction.	S and Te additions needed to regain grain growth capabilities.
Nb	α phase	• Increases Hc while decreases Br. • Forms precipitate with Carbon. • Detrimental for columnar grain growth. Carbon is a strong γ stabilizer and needs to be eliminated.	
Hf		• Enhances high-temperature properties. • Precipitates at the grain boundary and helps in improving creep properties.	

greatly influences the structure of these alloys which is responsible for superior soft magnetic properties in comparison to other soft magnetic alloys.

In Chapter 6, we have presented a case study on soft magnetic FINEMET-type alloys.

2.5 EXTRACTIVE/PROCESS METALLURGY

Figure 2.7 shows the schematic representation of extraction of iron from iron ore via various routes. In this book, we have included case studies on iron-making (blast furnace) and basic oxygen steelmaking (LD furnace).

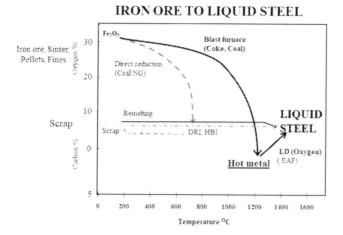

FIGURE 2.7   Iron ore to liquid steel: Schematic representation.

2.5.1   BLAST FURNACE IRON MAKING

A blast furnace is an efficient countercurrent heat exchange apparatus with a degree of heat utilization up to 85–90% (Biswas 1981, Smith 2017). With the advent of technology, many alternative routes of iron making have emerged in recent years, which utilize cheap grade materials and fuels to bring down the production cost. In spite of the high-energy requirements and environmental impact, the blast furnace route still dominates the steel industry, mainly due to higher production rate (up to about 10000–14000 THM/Day) and higher degree of heat utilization when compared to other alternative routes (Ghosh and Chatterjee 2015).

2.5.1.1   Hot Metal Production

In a blast furnace, raw materials like iron ore, sinter, coke, and limestone are charged from the top and preheated air is blown into the bottom. The raw materials require about 6–8 hours to descend to the bottom of the furnace where they become the final product of liquid hot metal and slag. These liquid products are drained from the furnace at regular intervals. The hot air that was blown into the bottom of the furnace ascends and exits from the top in 6–8 seconds after going through numerous chemical reactions.

2.5.1.2   Furnace Variables

2.5.1.2.1   Iron Ore

Iron is a 3-d element. Its melting point is 1539°C. It exists in nature mainly in the form of oxides, i.e. consists mainly of hematite (Fe_2O_3), which is red; magnetite (Fe_3O_4), which is black; limonite or bog-iron ore ($2Fe_2O_3 \cdot 3H_2O$), which is brown; and siderite ($FeCO_3$), which is pale brown. Haematite (Fe_2O_3) and magnetite (Fe_3O_4) are the most common types of ore. Deposits with less than 30% iron are commercially unattractive. Some ores contain as much as 66% iron, while the majority are in the 50–60% range. Iron ore's quality is also influenced by its other constituents, which are collectively known as gangue. These gangue materials are removed as slag. Silica (SiO_2) and phosphorus-bearing compounds (usually P_2O_5) are especially important as they affect the composition of the hot metal and pose extra problems in steelmaking. Iron ore fines are often agglomerated and used in form of sinter and pellets.

2.5.1.2.2 Coke (Biswas 1981, Smith 2017)

Metallurgical coke is produced by the destructive distillation of coal that is heated in the absence of oxygen in coke ovens at a high temperature (1100°C) in order to remove most of the volatile components associated with it. The final solid that is left over is called metallurgical coke and has a very low volatile content but the ash constituents remain encapsulated. A high-quality coke should be able to support a smooth descent of the burden. It should possess the lowest amount of associated impurities, highest thermal energy, highest metal reduction capacity, and optimum permeability for the flow of gaseous and molten products. Use of high-quality coke in a blast furnace will ensure a lower coke rate, higher productivity, and hence lower hot metal production cost.

2.5.1.2.3 Flux (Ghosh and Chatterjee 2015)

Fluxes are added to sinter or charged directly into the blast furnace in order to liquify ore and sinter gangue and fuel ash, thus converting them to free-flowing slag that can be run out of the furnace with ease. The limestone is the most popular flux in blast furnace and sintering process. The limestone charged in the blast furnace must be in lumps (25–60 mm across) and most important, it must be free from harmful sulphur, phosphorus, and silica.

2.5.1.3   Blast Furnace Process Description (Ghosh and Chatterjee 2015, Zhou et al. 2016)

Above the tuyeres, a blast furnace can be broadly classified into two zones (Biswas 1981):

- Indirect reduction zone and
- Direct reduction and melting zone

2.5.1.3.1 Indirect Reduction Zone

It can be further classified into the preheating zone (20°C to about 900°C) and the thermal reserve zone (about 900°C). Iron ore is reduced by a series of chemical reactions; these reactions can be described as the following:

$$3Fe_2O_3 + CO(or\ H_2) \rightarrow 2Fe_3O_4 + CO_2(or\ H_2O) \tag{2.25}$$

$$Fe_3O4 + CO(or\ H_2) \rightarrow 3FeO + CO_2(or\ H_2O) \tag{2.26}$$

$$FeO + CO(or\ H_2) \rightarrow Fe + CO_2(or\ H_2O) \tag{2.27}$$

Reaction (2.25) begins at about 450°C in the preheating zone. Reaction (2.26) occurs at about 600°C in the thermal reserve zone. Here, wustite (Fe_xO, a non-stoichiometric oxide of iron) is formed. As we descend down further, the thermal reserve zone thereafter consists of a chemical inactive zone followed by indirect reduction zone of wustite, where a reaction (2.27) occurs above 700°C. Some unreduced wustite is left over, which is reduced in the direct reduction zone.

2.5.1.3.2 Direct Reduction and Melting Zone

This zone is just above the tuyers; as a result, it is a high-temperature region. In this zone, the temperature varies from 900°C to about 2000°C at the tuyers.

The reactions occurring in this region can be described as follows:

$$FeO + CO(or\ H_2) \rightarrow Fe + CO_2(or\ H_2O) \tag{2.28}$$

$$FeO + C(or\ H_2) \rightarrow Fe + CO(or\ H_2O) \tag{2.29}$$

$$CaCO_3 \rightarrow CaO + CO_2 \tag{2.30}$$

$$CO_2 + C \rightarrow 2CO \tag{2.31}$$

$$MnO + C \rightarrow Mn + CO \tag{2.32}$$

$$P_2O_5 + 5C \rightarrow 2P + 5CO \tag{2.33}$$

$$SiO_2 + 2C \rightarrow Si + 2CO \tag{2.34}$$

$$CaO + SiO_2 \rightarrow CaSiO_3 \tag{2.35}$$

$$S + CaO + C \rightarrow CaS + CO \tag{2.36}$$

$$C + O_2 \rightarrow 2CO \tag{2.37}$$

$$C + H_2O(Steam) \rightarrow CO + H_2 \tag{2.38}$$

Reaction (2.28) is known as indirect reduction and it is exothermic in nature. Reaction (2.29), commonly known as direct reduction, is endothermic, but it requires less carbon in comparison to indirect reaction. The limestone descends in the blast furnace and remains solid while going through the decomposition, as shown in Reaction (2.30) at about 875°C. The reaction product CaO is used to remove impurities like sulphur and silica, as shown in Reaction (2.35) and (2.36). Reactions (2.32), (2.33), and (2.34) show reduction of other oxides that go into the hot metal as impurities.

Blast (preheated air from stoves along with oxygen and humidity additions) is blown in through the tuyeres. In order to decrease the coke rate, some other form of fuel is also added through a lance in the tuyere. Carbon in the form of coke is burnt at the tuyers according to Reaction (2.37) to produce carbon monoxide.

This reaction is highly exothermic in nature and is responsible for the energy re-quirements of the blast furnace. The burning area just inside the tuyeres is usually mentioned as the raceway. Since the gas temperature in the raceway is high (usually about 2000°C), and hence, if carbon is present in excess, carbon dioxide (CO_2) from other reactions will react with carbon-forming carbon monoxide (CO), as shown in Reaction (2.31). It is also known as Boudouard or solution loss or carbon gasifi-cation reaction. Boudouard's reaction is important to some extent as it reduces the amount of reducing agents needed in the blast furnace. It is highly endothermic; hence, it consumes a large amount of energy and also deteriorates CO utilization. If the reaction rate is too high, the fuel consumption will increase. In almost all si-tuations, the latter case dominates over the former case. Hence, in order to decrease the fuel rate in a blast furnace, it is very important to decrease the reaction rate of Boudouard's reaction. Steam going into the flame will crack and form hydrogen gas (H_2), as shown in Reaction (2.38). It is also known as a water gas shift reaction. It is similar to Boudouard reaction. It generates carbon monoxide and hydrogen, both of which are used as reducing agent. This reaction is endothermic in nature. Carbon monoxide and hydrogen gas will then pass up through the shaft and act as reduction agents of the iron oxides. Nitrogen gas, hydrogen gas, and carbon monoxide leave the raceway with high temperature. During the passage up in the furnace the gas temperature decreases as heat is used to heat the burden material and also in other endothermic reactions. In this part of the blast furnace, the iron will soften and melt, leading to the formation of cohesive zone. After melting, liquid iron drips through the active coke layer and the deadman zone. The active coke layer is the coke that flows towards the raceway, where most of the coke is burnt. The deadman zone consists of coke that does not flow towards the raceway and it can be located inside the blast furnace for a long time. It will mainly be consumed by dissolving into the liquid iron and also to some extent in reaction with other species (e.g. MnO, P_2O_5 and SiO_2) as shown in Reactions (2.32), (2.33), and (2.34).

The products that we get are hot metal and slag, which are tapped down at regular intervals from the bottom of the furnace. Hot metal typically contains 4–4.5% carbon, 0.6–1.5% silicon, 0.03–0.05% sulphur, 0.7–0.8% manganese, and 0.15% phosphorus. Tapping temperatures are in the range 1400–1500°C. In order to save energy, the hot metal is transferred directly to the steel plant with a tem-perature loss of about 100°C.

Blast furnace slag typically contains about 35% SiO_2, 20% Al_2O_3, 32% CaO, and about 8–10% MgO with traces of P_2O_5, K_2O, etc.

2.5.1.4 Productivity

The term *productivity* can be coined in order to get an idea about the performance of the furnace in comparison with other furnaces elsewhere in the world. In one of the earliest approaches, the production rate, (Biswas 1981), also known as smelting rate or output, is denoted in equation (2.39) as

$$P = \frac{Q}{K} \tag{2.39}$$

Where, P = Productivity, THM/Day, Q = Coke burned, Tonnes/Day, K = Coke consumed, Tons/THM

In recent years, due to rise in usage of various auxiliary fuels (as injectants) like pulverised coal, gas, oil, steam, plastic, etc., a more generalized approach was needed in order to get a clear idea regarding the performance index of blast furnace. This can be done by bringing the idea of working volume or useful volume that is the furnace volume calculated from the height taken from the tuyere axis to one meter below the bell in open position. In this way, productivity is estimated as given in equation (2.40):

$$P = \frac{\text{Production in THM per day}}{\text{Working Volume in m}^3} \tag{2.40}$$

Where, P = Productivity, (THM/Day)/ Working Volume (m^3).

2.5.2 STEELMAKING

Hot metal from the blast furnace is further processed. This is done in primary steelmaking in a basic oxygen furnace (BOF) or an electric arc furnace (EAF). Modern BOF and EAF have a capacity of about 400 tons and a tap to tap time of about 40 minutes. The rest of the steelmaking process involves further refining that involves removal of impurities and fine tuning of the composition in order to meet the customer requirements.

2.5.2.1 Basic Oxygen Furnace (BOF): LD (Linz Donawitz) Furnace (Turkdogan 1996)

The purpose of the basic oxygen steelmaking (BOS) process is to refine the hot metal produced in the blast furnace into raw liquid steel. Basic oxygen steelmaking (BOS) is the most widely used process for producing crude steel from molten hot metal. The process involves blowing oxygen through hot metal to reduce its carbon content by oxidation. There exist many different types of furnaces, or converters. These are generally divided into top-blowing, bottom-blowing, and mixed blowing. Top blowing through a vertical lance was invented in 1952–1953 at Linz and Donawitz in Austria and the process is known as LD, or BOP (basic oxygen process) (Trenkler 1960). The main functions of the basic oxygen furnace (BOF) are decarburization and removal of phosphorus from the hot metal. Exothermal oxidation reactions generate a large amount of heat energy. This amount of heat is higher than what is required to reach the target temperature of the steel. The additional heat is used to melt scrap metal and/or iron ore additions (Penz 2018).

2.5.2.1.1 Input: Raw Materials Charged into the Converter

Raw materials include the following: (Ghosh and Chatterjee 2015, Turkdogan 1996)

- Liquid hot metal from the blast furnace after specific pretreatments such as desulfurization or dephosphoration.

- Other iron-based additions, principally scrap and ore.
- Additions required to form a slag of appropriate composition, consisting primarily of lime (CaO) and dolomitic lime (CaO-MgO), usually in lumps of 20 to 40 mm.
- Pure oxygen injected either through lance from the top or through bottom tuyeres.

2.5.2.1.2 Output: Materials Produced After Blowing Operation Is Complete

- Liquid steel
- Exhaust emissions contain approximately 80–90% CO. It is retrieved by the enclosed or suppressed combustion hood. This gas is rich in CO and is commonly used in furnace burners CO is carbon monoxide.
- The slag is poured off the vessel at the end. Gases and slags are valuable and are properly collected and stored.

2.5.2.1.3 Reactions in BOF (Ghosh and Chatterjee 2015)

The hot metal charge is refined by rapid oxidation reactions on contact with the injected oxygen. Carbon in the hot metal is oxidized to CO (equation 2.41).

$$C + \tfrac{1}{2}\,O \rightarrow CO \qquad\qquad (2.41)$$

CO is partially oxidized into CO_2 above the melt (post-combustion). These gaseous reaction products are discharged via the exhaust hood. The ratio $CO_2/(CO+CO_2)$ is called a post-combustion ratio.

Other oxidation reactions during refining include the following (2.42 to 2.46):

$$Si + O_2 \rightarrow SiO_2 \qquad\qquad (2.42)$$

$$2P + (5/2)O2 \rightarrow P_2O_5 \qquad\qquad (2.43)$$

$$Mn + \tfrac{1}{2}O_2 \rightarrow MnO \qquad\qquad (2.44)$$

$$Fe + \tfrac{1}{2}O2 \rightarrow FeO \qquad\qquad (2.45)$$

$$2Fe + (3/2)O_2 \rightarrow Fe_2O_3 \qquad\qquad (2.46)$$

These oxides combine with the other oxides (lime, dolomite lime) to form a liquid slag that floats on the surface of the metal bath.

2.5.2.1.3.1 Reactions at Oxygen Jet Impact for Top-Blown Converters

On impact, the components are oxidized in proportion to their hot metal content (Fe, C, Si, etc.). These primary oxidation products are then ejected, either mostly through the metal (penetrating jet, low lance) or to the slag (soft blow, high lance).

Any FeO in droplets going through the metal is reduced by C, only partially when the Si content is high (beginning of the blow) or when the carbon content is low (end of the blow). FeO in droplets is ejected directly onto the slag and increases its FeO content. This foamy slag can react violently with metal droplets leading to slopping.

2.5.2.1.3.2 Removal of Impurity During Blowing Operation　Decarburization is the most extensive and important reaction during oxygen steelmaking. About 4.5 weight % carbons in the hot metal is oxidized to CO and CO_2 during the oxygen blow, and steel with less than 0.1 weight % carbon is produced.

2.5.2.1.4 Slag

The slag contains oxides arising from some oxidation reactions (SiO_2, P_2O_5, FeO, and MnO), added fluxes (CaO, MgO) and refractory wear (MgO).

2.6　NICKEL-BASE SUPERALLOYS

Nickel-based superalloys are used for high-temperature applications on turbine blades and are the most stable superalloys at high temperatures. From application point of view, stress to rupture and time to rupture are conflicting with each other. Any improvement in either of these will lead to the deterioration of the other property (Pollock and Tin 2006, Tancret 2012).

Nickel-based super alloys are used in a wide range of applications. Applications include high-pressure turbine, aerospace, and nuclear industry due to their superior resistance to oxidation, corrosion, creep, and fatigue at elevated temperatures (Pollock and Tin 2006). Multi-component nickel-based superalloys can contain over 10 elements. Each of the alloying elements is important for a specific purpose. While the same elements may cause precipitations of detrimental phases that must be avoided to achieve the desired properties. Because of its importance, a great deal of research has been done on the physical metallurgy of Ni-based superalloys.

2.6.1　CRITICAL PHASES

In metallurgical terms, nickel-based superalloys include the following four types of phases: (Tancret 2012)

1. Disordered FCC_A1 or austenitic FCC phase (γ): It is a non-magnetic phase. All nickel-based superalloys contain this phase as matrix phase. Gamma (γ) phase is stabilized by Ni, Co, Fe, Cr, Ru, Mo, Re, and W, as its atomic radii are close to Ni; thus, it partitions to gamma (γ) phase and stabilizes it.
2. Geometrically Close Packed (GCP) phase: Ordered FCC_L1$_2$ phase or gamma prime (γ') phase is the main strengthening phase, which is responsible for superior high-temperature mechanical properties in nickel-based super-alloys. In the presence of a substantial amount of Nb, another strengthening phase, the BCT D022 (γ'') phase, appears.

a. Gamma prime (γ'): Gamma prime (γ') phase is stabilized by Al, Ta and Ti, as its atomic radius is larger than Ni. Thus, it partitions to gamma prime (γ') phase and stabilizes it.

b. BCT D022 (γ'') phase: It is stabilized by Nb and provides very high strength for low to intermediate temperatures. BCT D022 (γ'') phase is unstable above 649°C.

3. Carbides: Carbon combines with elements like Ti, Ta, Hf, Nb to form metal carbides MC, $M_{23}C_6$, M_6C. MC usually decomposes during heat treatment to form $M_{23}C_6$ and M_6C at the grain boundaries.

4. Tetragonally Close Packed (TCP) phase: Co, Mo, W, Cr, etc. are added for stabilizing gamma (γ) phase, but these elements are also responsible for the precipitation of detrimental TCP phases like σ, R, P, μ, Laves. TCP phases nucleate and grow and deplete the γ-matrix from solid solution strengtheners resulting in decreased creep resistance of nickel-based superalloys. TCP phases can be suppressed by adding ruthenium, while σ phase alone can be suppressed by the adjusting/optimizing alloy chemistry. TCP phases are not always detrimental. But the presence of TCP phases in any amount that can be considered as more than trace amounts can result in a decrease in ductility and can affect mechanical and corrosion properties.

2.6.2 Effect of Alloying Elements

Ni-based superalloys can contain up to 20 elements (Egorov-Yegorov and Dulikravich 2005, Jha et al. 2014, Pollock and Tin 2006):

- Co, Cr, Fe, Mo, W, Ta, Re enhances solid solution strengthening.
- Cr, Ti, Nb, Hf, Mo, W, Ta enhances the grain boundary strengthening with carbides.
- Al, Ti enhances the elevated-temperature strength and resistance to creep deformation.
- Al, Cr, Y, Ce, La improves the oxidation resistance.
- Cr, Co, Si, Th, La improve hot corrosion resistance.
- B, C, Zr, Hf act as grain boundary strengthening (refining).

2.7 ALUMINUM ALLOYS

Aluminum is one of the most abundant elements in the Earth's crust. Aluminium alloys are vulnerable to corrosion. Researchers around the globe have been working on designing new chemistry as well as the manufacture protocol (heat-treatment) that will provide with an alloy with improved mechanical properties and corrosion resistance at temperatures in the vicinity of 400°C. Scandium alone as well as scandium along with zirconium improves mechanical strength and corrosion resistance of aluminum alloys, but Sc addition is expensive. Depending on size and product, the cost of the end product may increase three to four times for the addition of Sc by about 0.2 weight %. Aluminum alloys with scandium have been used in aircraft (military applications) in Russia (Røyset and Ryum 2005). In the United

States, Al-Sc-based aluminum alloys have been used in baseball bats. It was launched in 1997 by Easton with Kaiser Aluminum and Ashurst Technology, and then by ALCOA. Smith and Wesson manufactured frames for a series of handguns using Al-Sc alloys (Røyset and Ryum 2005).

Al_3Sc is a thermodynamically stable phase and precipitates at about 300–400°C, while other precipitation hardening phases in aluminum alloys precipitate at about 160–200°C. This makes the Sc addition a complex process (Dorin et al. 2019, Thermo-Calc Software TCAL5 2018, Røyset and Ryum 2005). Initial classifications of 2XXX, 6XXX, and 7XXX were based on particular alloying elements like 2XXX is Al-Cu based, 6XXX is Al-Mg-Si based, while 7XXX is Al-Mg-Zn based (Røyset and Ryum 2005). However, most of the alloys in use contain several alloying components. Most metastable phases are thermodynamically unstable, but may be found in the final microstructure (De Luca et al. 2018, Dorin et al. 2019, 2017, Røyset and Ryum 2005). The addition of scandium improves the nucleation and stability of these metastable phases (Dorin et al. 2017, Gao et al. 2020, Mondol et al. 2017). Therefore, it is important to study various stable and metastable phases in aluminum-based alloys to minimize Sc addition, while designing a new chemistry and manufacturing protocol. Aluminum alloys have been studied under the framework of CALPHAD approach (Andersson et al. 2002, Du et al. 2018, Sarafoglou et al. 2019, Tang et al. 2018, Thermo-Calc Software TCAL5 2018, Thermo-Calc Software MOBAL4 2018, Zhang et al. 2019). Recently, CALPHAD-based calculations have been combined with AI-based algorithms to design a new Al-based alloy composition (Jha and Dulikravich 2020,Zhang et al. 2019).

Precipitation sequence of critical phases in aluminum-based alloys are as follows:

- 2XXX: *Supersaturated solid solution* → *GP-zones* → θ'' → θ *(Al₂Cu) phase.* θ (Al2Cu) phase is a stable phase. This class of alloys achieves superior strength when θ'' and θ' are the predominant strengthening phase (Røyset and Ryum 2005).
- 6XXX: *Supersaturated solid solution* →*GP-zones* → β'' → β *(Mg₂Si).* β (Mg2Si) phase is a stable phase, β'' is the predominant phase. β'' is observed after aging (Røyset and Ryum 2005).
- 7XXX: *Supersaturated solid solution* →*GP-zones* → η' → η *(Mg₂Zn).*

The η (Mg₂Zn) phase is a stable phase. This class of alloys achieves superior strength when η' and η phases are predominant while maximum hardness is achieved when η' is predominant (Røyset and Ryum 2005).

2.8 TITANIUM ALLOYS

2.8.1 HIGH-TEMPERATURE APPLICATIONS

Titanium alloys are a very important structural material for the aerospace industry, and currently around 80% of global titanium is consumed by this sector (Polmear et al. 2017). Titanium alloys exhibit high strength, fracture toughness, creep strength, excellent resistance to corrosion, and an enhanced fatigue life when compared with other structural materials. Carbon-epoxy composites possess the highest specific strength. Apart from carbon-epoxy composites, titanium alloy has the highest specific strength among the structural materials used in aerospace industry. Density of titanium alloy (4.5 g/cm^3) is lower than steel- and nickel-based alloys, but higher when compared to aluminum alloys (2.7 g/cm^3) and carbon-epoxy composites (1.5–2 g/cm^3). Extraction of titanium is complex and expensive process. In the aerospace sector, it finds application mostly in military aircrafts. Currently, it is used for airframe structures, landing gear components, and jet engine parts. It is used in turbine blades and discs because it maintains structural integrity at high temperatures. The manufacture of aircraft components require a specialized material removal process, which adds to the overall cost of the component (Polmear et al. 2017).

The first commercial use of alloys based on the Ti-Al-Cr-V system was as a β-Ti alloy in Lockheed SR-71 Blackbird military aircraft. This alloy contained 13% vanadium, 11% chromium, 3% aluminum, and the rest was titanium. It was used for manufacturing the following parts of the SR-71 Blackbird: airframe, fuselage frame, wing, longerons, bulkheads, ribs, landing gear, and body skins. The SR-71 still has the fastest-breathing manned aircraft record since 1976 and was designed to fly above Mach 2.5. At Mach 2.5, aerodynamic friction and shock waves heated the skin to a temperature of about 300°C. Therefore, titanium alloys were used in the SR-71 because titanium has a higher strength at high temperatures compared to aluminum alloys (Polmear et al. 2017).

2.8.1.1 Phases

Alloys containing Ni, Mg, and Al as the base element, have a single crystal stable at the room temperature. But, titanium base alloys can have two stable allotropes, hexagonal close-packed (HCP) known as α-Ti and body centered-cubic (BCC) known as β-Ti. The chemistry of these two phases is virtually identical, but these phases have different properties. The α-Ti phase has low to medium strength, good toughness and ductility, is weldable, and has excellent creep resistance at elevated temperatures. The β-Ti phase is associated with high strength and fatigue, is fully heat treatable, is creep resistant at intermediate temperatures, and is less ductile than α-Ti (Polmear et al. 2017).

Titanium can exhibit a valency of two, three, or four. Elements at a lower valency state when compared to titanium will promote stabilization of α phase. While the elements at a higher valency state will promote stabilization of β phase. Aluminum and tin have been used for stabilizing the α-phase, while Cr, V, Mo, Nb, and Fe have been used for stabilizing the β-phase (Polmear et al. 2017). Elements like carbon, nitrogen, silicon, and rhenium are neutral elements and are added for improving various properties. Thus, the thermal treatment and chemical composition significantly affect the stability and the desired amount of phase needed in the alloy for application at a given temperature (Polmear et al. 2017). Additionally, there exists phases like titanium aluminides, Ti$_3$Al (D019), and γ-TiAl, which are also quite important for the aerospace sector (Gupta and Pant 2018, Jha and Dulikravich 2019, Polmear et al. 2017, Wu 2006).

At about 9% aluminum, the α-phase and the titanium aluminides phases are both stable (Jha and Dulikravich 2019, Gupta and Pant 2018, Wu 2006). Commercially used titanium alloys are mostly developed as (α + β)-Titanium alloy. In (α + β)-Titanium alloy, titanium aluminides are considered as detrimental phase as it is brittle (Polmear et al. 2017).

In titanium alloys, there exists a ω-phase, which is detrimental for all types of titanium-based alloys. Be it for aerospace application, in shape-memory alloys, or for biomedical application, the ω-phase is detrimental for all of them. Attempts are being made to develop ω-free alloys for all of these applications. It is not stable at room temperature, but the ω-phase has been observed in the final microstructures (Ferrari et al. 2019, Polmear et al. 2017).

Thus, a user must focus on chemistry and heat-treatment protocol while manufacturing these alloys. In this work, we presented two case studies for titanium alloys.

2.8.1.2   Titanium Aluminides (Gupta and Pant 2018, Wu 2006)

Titanium aluminides have the lowest densities among various aluminides (titanium, iron, nickel) used for aerospace applications. Titanium aluminides that are of interest are as follows:

1. Ti_3Al with D019 structure (also known as $α_2$) and ordered face-centered tetragonal $L1_0$ structure known as γ-TiAl. Ti3Al melts at 1600°C and is used for applications up to 700°C. It possesses poor oxidation resistance, but its presence in small amount along with γ-TiAl helps in improving the ductility of γ-TiAl alloys.

2. γ-TiAl melts at 1460°C and is used for applications up to 1000°C. It possesses fair, oxidation resistance, but in its pure form it is brittle, but it remains in the ordered form until the melting point. It has the lowest density among all the aluminides at about 3.4 gm/cc, while the density of alloys containing γ-TiAl is about 3.4–3.9 gm/cc. Researchers around the globe have been working on designing new compositions and manufacturing protocols that will be helpful in improving the ductility and high temperature oxidation resistance of γ-TiAl. γ-TiAl were applied first by Mitsubishi in 1999 in the automotive sector as turbocharger rotors. In the aerospace sector, General Electric used it as alloy GE48-2-2 (Ti-48Al-2Nb-2Cr at%) in its aircraft engine GEnx-1B used in the Boeing 787 Dreamliner (Polmear et al. 2017). Due to these recent developments, rather than remaining confined to the automotive sector, γ-TiAl is seen as a contender among materials that can be used in future in the structure of hypersonic trans-atmospheric aircraft. These aircraft are expected to suffer severe aerodynamic heating during ascent and decent and thus materials that will be used for the structure of the body of the aircraft are expected to be stable in the vicinity of 1000°C (Gupta and Pant 2018).

As mentioned before, pure γ-TiAl is brittle and possess fair oxidation resistance. Ti_3Al phase helps in improving ductility, but it is comparatively poor in oxidation resistance. For application in the structure of hypersonic trans-atmospheric aircraft, one has to work on improving the oxidation resistance of γ-TiAl alloys.

In order to improve upon various properties for desired applications, a few elements are added along with titanium and aluminum (Gupta and Pant 2018, Wu 2006).

Various elements and their role in titanium aluminides are as follows:

1. Cr, V, Mn, and Si: Helps in improving oxidation resistance, but it reduces oxidation resistance.
2. Nb, Ta, Mo, and W: It improves oxidation resistance, but these are heavy elements and can increase the density of the alloy if added in large amount.
3. Si, C, and N: These are added in small amounts. It helps in improving the creep resistance of the alloys.
4. Ni: Improves ductility.
5. W, V, Mn, and Cr: Reduces grain size of γ-TiAl phase and also α_2-Ti$_3$Al phase. This helps in improving ductility.
6. C and B: In small amounts, helps in reducing the grain size.
7. C: If Carbon is present in excess of 1,000 weight ppm, it leads to precipitation of Ti$_2$AlC, which improves the creep resistance of the alloy.
8. B: It improves strength at the room temperature and at high temperature. If present in excess, it forms brittle TiB$_2$ phase.
9. W: It also improves creep resistance.
10. Ta: It is beneficial for the refining process, mechanical properties, oxidation, and creep resistance.
11. Mn and Cr: Improves ductility, reduces the volume fraction of the α_2-Ti$_3$Al phase, and refines grain size.
12. Mo: It improves tensile ductility more effectively than Cr and Mn addition.
13. Nb: Improves oxidation resistance and strength.
14. Re: Improves creep resistance.

From the previous description, it can be seen that elements added affect various properties of aluminides in various ways. Each of these elements can improve a specific property that is important for the application, but at the same time it can also deteriorate another property which is equally important. Thus, choosing an alloying element to design chemistry for a specific application is very critical in titanium aluminides. In order to improve upon ductility, it has been shown that α_2-Ti$_3$Al phase up to 10% volume can be helpful in addition to γ-TiAl phase.

2.8.2 BIOMATERIALS

The first joint replacement can be traced back to 1891 (Gluck 2011, Gudkov et al. 2020). Titanium alloys have been widely adopted for biomedical applications, particularly in implants (Gudkov et al. 2020, Long and Rack 1998). Titanium alloys are well known for biocompatibility and corrosion resistance (Marker 2017). For biomedical applications, it is desired that the alloy has a fully stabilized β-phase (Marker 2017). Titanium alloys with a fully stabilized β-phase possess comparatively lower Young's modulus (YM), which is important for an implant (Mohammed et al. 2014). Young's modulus of bone is about 10–40 GPA, while Young's modulus of common implants, including Ti-based implants can vary between 100–230 GPA. This difference will cause stress shielding in the bones as a result of non-uniform distribution of stress in the area around the implant. This can cause fractures of bones or can be considered as a failure of an implant. Additionally, lots of alloying elements (Al, V) are toxic to the human body (Boyce et al. 1992, Domingo 2002, Ito et al. 1995).

Thermodynamically, the β-phase can transform into the ω-phase during heat treatment (Aleixo et al. 2008, Mantri et al. 2019). The α-phase is a stable phase and for biomedical applications, it is desired that titanium-based alloy is free from the α-, α''-, and ω-phase. Additionally, the alloying elements must be analyzed for toxicity.

2.9 UTILIZING DATA FROM OTHER APPROACHES

2.9.1 Density Functional Theory/Ab-initio Method

DFT-based tools like VASP, LAMMPS, etc. generate lots of data that can be efficiently utilized through the application of AI algorithms.

2.9.1.1 VASP

In this book, we have utilized data generated through VASP in Chapter 6.

2.9.1.2 Schrodinger Materials Design Suite

A case study on designing refrigerants through AI-based algorithm and atomistic simulations have been presented in Chapter 3 (Bochevarov et al. 2013).

2.10 CASE STUDY # 2: INVERSE DETERMINATION OF PHASE-FIELD MODEL PARAMETERS FOR A TARGETED GRAIN SIZE THROUGH AI ALGORITHMS

2.10.1 Phase-Field Approach (Moelans et al. 2015, Chang and Moelans 2015)

In this section, we have provided a short background on phase-field approach. It is followed by a case study on AI application of phase-field data for optimizing grain size.

2.10.1.1 Background

Grain size plays a key role in determining the mechanical properties of metallic alloys. Smith and Zener's work introduced a means of controlling grain size by introducing a second-phase particle. This method, known as Zener pinning, works by pinning the grain boundaries on the particle-grain interface. Zener pinning is one of the benchmarks adopted by researchers working with controlling grain size in an alloy. Industrial examples of Zener pinning include high strength low alloyed (HSLA) steels and super alloys.

Zener pinning is a complicated phenomenon due to the interaction of various parameters. Parameters may include second phase particle size, morphology, volume fraction, coherency, anisotropic interfacial energy, and coarsening of the second phase particle. Consequently, only the most basic cases can be described analytically.

Computer simulations were therefore attempted to understand the interactions of the above-mentioned parameters and to obtain valuable information about the Zener pinning. Zener pinning has been studied through different computational techniques such as Monte-Carlo Potts models, front-tracking models, and phase-field models.

The phase-field model is a diffuse interface model where the evolution of arbitrary and complex grain morphologies can be investigated without any presumption on their form or mutual distribution. Because of these advantages, the phase-field model has been widely used to understand the Zener pinning phenomenon.

To understand this complex Zener pinning phenomenon, we used a new technique coupled with phase-field and machine learning. The effect of volume fraction, shape and size of second phase particle of Zener pinning drag have been studied through phase-field model individually in the past. This is the first work of its kind where all of these parameters are being analyzed together. In practical applications, it is possible to vary these parameters together and sometimes they are impossible to control precisely individually. Therefore, it is important to understand how the grain structure changes when one, two, or all of these parameters are operating concurrently.

2.10.1.2 Phase-Field Variables Identification for This Study

In this study, we obtained data from phase-field simulations where all parameters were varied. This has been instrumental in producing a substantial amount of simulation results that can be efficiently analyzed using AI algorithms. Artificial intelligence–based algorithms are suitable for non-linear problems and are known to determine complex interrelationships between design variables and target properties. Additionally, we used the concept of parallel coordinate chart (PCC) as a visualization tool to determine the parameters that will be useful to control grain size. This will be helpful in addressing the requirements of the materials (with respect to microstructure) that is to be deployed in a particular application.

In Zener's pinning, second-phase particles have a pinning effect on grain boundaries and affect the mobility of grain boundaries, ultimately stopping grain growth. The pinning pressure (P_z) exerted by a second-phase particles, characterized by particle diameter (P_d) and the volume fraction (V_f) of the particles, is given by

$$P_z = \frac{3V_f \gamma_{bd}}{2P_d} \tag{2.47}$$

When the driving pressure for grain growth is equal to the pinning pressure of the particles, the grain growth seizes and a limiting grain size ($\overline{D_{lim}}$) is reached.

$$\frac{\alpha \gamma_{bd}}{\overline{D_{lim}}} = \frac{3V_f \gamma_{bd}}{2P_d} \tag{2.48}$$

$$\overline{D_{lim}} = \frac{2P_d \alpha}{3V_f} \tag{2.49}$$

α is second-phase particle shape factor.

Although the assumptions Zener made are simplistic and known to under-estimate the pinning effect of a particle distribution, this relation is still used frequently in practical applications for prediction of the limiting mean grain size. Chang and Moelans (2015) have shown through phase-field modeling that for the same volume fraction smaller particles and needle shaped particles are more effective in pinning the grain boundaries. Vanherpe et al. (2007) demonstrated that the pinning effect of a particle is greater to increase the volume fraction and aspect ratio. The total fraction of particles presents at boundaries increases with higher volume fraction and with aspect ratio. This contributes to the greater pinning effect. A mathematical expression for the limiting grain size and particle size and aspect ratio and volume fraction is presented in equation 2.50.

$$\overline{D_{lim}} = K \frac{l}{1 + ar_a} \frac{1}{V_f^b} \tag{2.50}$$

Here, l is the long axis size of the particle; r_a is the aspect ratio; and K, a, and b are fitting parameters.

Variables that influence the Zener pinning can be summarized in Table 2.3

2.10.1.3 Application of AI Algorithms

It is important to analyze any database using statistical toolboxes prior to developing mathematical models or applying AI-based algorithms. We determined the relative contribution of all the input variables on volume fraction. Calculations were performed by means of a Smoothing Spline ANOVA algorithm in ESTECO modeFRONTIER software (Table 2.3).

2.10.1.4 Application of AI-Based Algorithm on Phase-Field Data

From the phase-field model, we were able to generate a significant amount of data that can be useful for providing meaningful insights for the system under consideration. First of all, we made an attempt to find a correlation between the input variables (for a phase-field model) and the final grain diameter. Figure 2.8 shows a graphical representation of the contribution of various input parameters on the final grain diameter.

To develop a model, we need to train the model over a given data set and then test it over data that was not exposed to the model during training. In this work, we created three sets of training and testing sets, and the models were trained on all the training set and the whole data set. We used several supervised machine learning algorithms and compared their performance by testing it on the mentioned testing sets. Thereafter, we chose the algorithm that performed best on all the testing sets and used the model created by that algorithm while training over the complete data set. Two algorithms performed consistently better: Anisotropic Kriging and k-Nearest Neighbor algorithm. Hence, we focused on these two algorithms and finally chose the k-Nearest Neighbor algorithm with 11 points, where k = 11.

At this point, we demonstrated that machine learning algorithms can be efficiently used to develop predictive models that can mimic the outcomes of the phase-field models, where the execution time of metamodels is several orders of magnitude lower when compared with the execution time of phase-field models.

TABLE 2.3
Phase-field Model Variables and Its Contribution to
Grain Size Estimated Through ANOVA in modeFRONTIER

Model Variables	Contribution Index
Particle size or diameter (P_d)	0.0751
Particle volume fraction (V_f)	0.236
Eccentricity or particle aspect ratio (ϵ)	0.452
Diffusivity or mobility (M)	0.184
Surface to grain boundary energy ratio $\left(\dfrac{\gamma_{surf}}{\gamma_{bd}}\right)$	0.0531
Mean absolute error	1.815
Mean relative error	0.110
Mean normalized error	0.045
R-squared	0.914

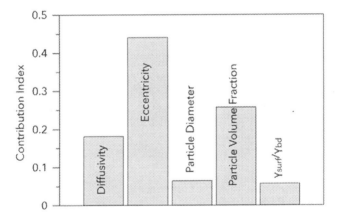

FIGURE 2.8   Relative contribution of phase-field model parameters on final grain diameter.

2.10.1.4.1 Generation of New Input Parameters

During analysis of the given data set, we observed that there are regions in the variable space where the distribution of input parameters was scarce. After developing the metamodels, it is possible to predict the final grain diameter for any combination of variable parameters within the variable bounds over which the model is trained. Hence, we generated new input parameters by application of Sobol's algorithm, as it is well known for its ability to generate random sequences for multi-dimensional hyperspace. In this work, we generated 10,000 new sets of parameters that cover the entire hyperspace under study. Execution of this large set of parameters through phase-field code is not feasible, but this information will be

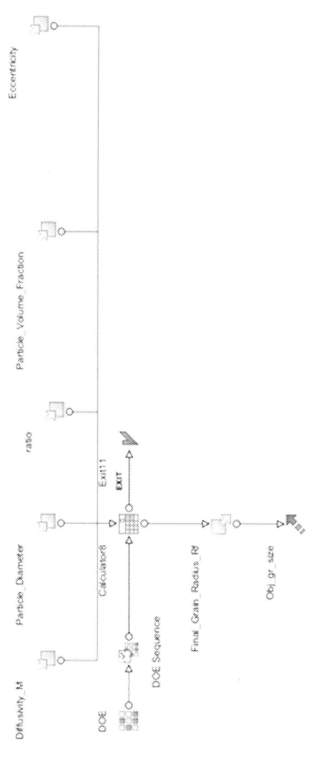

FIGURE 2.9 Workflow from modeFRONTIER for Case Study # 2.

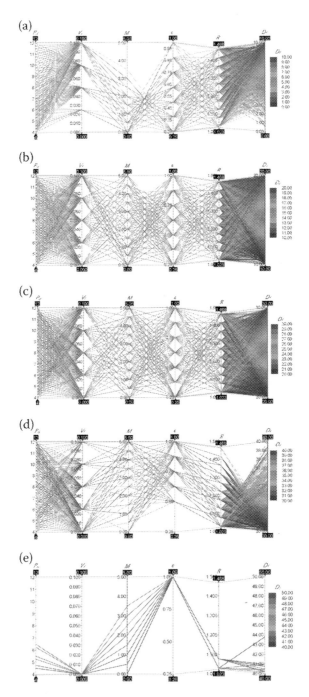

FIGURE 2.10 Parallel coordinate chart (PCC) representing all the phase-field model parameters and it's effect on final grain size.

helpful for studying the system. Thus, we used the selected metamodel as a prediction tool to predict the final grain diameter for the new parameters. Figure 2.9 shows the workflow developed in modeFRONTIER.

2.10.1.5   Parallel Coordinate Chart (PCC)

In the previous section, we demonstrated the efficacy of developing metamodels and were able to use it as a prediction tool for predicting the final grain diameter for a significantly large set of input parameters within a few minutes. Now, we proceeded to our next work that is to use all of this data to draw meaningful conclusions. We used the concept of parallel coordinates chart, which is an effective tool to analyze and visualize multi-dimensional and large data set.

In materials design, final grain size significantly affects the bulk properties of the material. Grain size affects the bulk properties differently, so it is important for a computational materials scientist to provide guidelines regarding the influence of input parameters on the variation of the final grain size. Hence, in this work, we divided the final grain diameter in five parts between 0 and 50 nm and plotted the parallel coordinate charts for these bounds. This way, it will be helpful for choosing a set of input parameters that seems to be most suitable as per the expertise of a researcher working on phase-field modeling.

Figure 2.10 shows the correlation between various input parameters of the phase-field model if we want to achieve a final grain size between 1 and 10 nm (a), 10 and 20 nm (b), 20 and 30 nm (c), 30 and 40 nm (d) and between 40 and 50 nm (e).

2.11   Conclusions

Applications of an AI-based approach proved to be helpful in determining phase-field model parameters for a targeted grain size. One can analyze thousands of different combinations of phase-field model parameters and determine their effect on the final grain size in a few minutes through the present approach. Performing such an analysis under the framework of phase-field modeling is not feasible due to computational cost. New parameters can be included in these AI-based models, and the entire work can be performed again within a matter of a few hours.

3 Artificial Intelligence Algorithms

Essentially all models are wrong, but some are useful.

G.E.P. Box

3.1 INTRODUCTION

In the AI/ML community, all of the results are generated through mathematical equations. We use terms like *random number*, but it is a sequence that can be reproduced if the user wants to get the same sequence. Similarly, in artificial neural networks (ANNs), inputs are associated with a weight (bias) in the hidden layer, and are processed through a mathematical equation to generate output values. Iterations are performed to reduce the error associated with a model. In optimization through evolutionary or genetic algorithms, a random population is generated at the beginning of optimization process. But, each member of the random population is a potential solution or a computer program or a mathematical model. Thus, nothing is random in random number generation, and nothing is hidden in the hidden layer of the ANN.

In this chapter, we explain various concepts of artificial intelligence in a balanced way. In this book, we have included case studies where data were generated or acquired from different sources and environments. It is important to process this data through several concepts of artificial intelligence and statistics. In this chapter, several AI-based approaches have been explained that will be helpful for the reader in following the case studies.

3.2 DATA MINING

Data mining is the art of extracting valuable insights from a given data set. Valuable information can include determining various patterns and correlations to the development of predictive models that can be used in the future. In order to process given data, a user can refer to a certain statistical approach, or algorithms based on concepts of artificial intelligence. Any statistical or AI-based approach will provide insights based on the functioning of the algorithm. Thus, it still depends on the user and their group, the information they consider acceptable, and the approach they consider reliable.

An experienced data scientist may be able to propose a set of algorithms that can work for a given test case and save a lot of time, but following any random approach may be misleading. Thus, it is always a good habit to analyze and preprocess a data set before drawing conclusions based on mere intuition.

DOI: 10.1201/9781003167372-3

3.2.1 Uncertainties

Uncertainty quantification is one of the major challenges faced by researchers in the MGI/ICME community. A data scientist must determine the sources of uncertainty and subsequently work on the development of mathematical models to address them. Broadly, uncertainty can be categorized as (Panchal et al. 2013):

- Aleatory or irreducible uncertainty: Its source is randomness of materials. It may be addressed through probability theory.
- Epistemic or reducible uncertainty: Human error gives rise to epistemic uncertainty. This may be due to lack of familiarity with process, system, assumptions, or approximations, etc. Researchers tackled epistemic uncertainty, but with limited success through the Bayes probability theory.

3.2.1.1 Uncertainty Propagation

The design of materials and alloys involves working across multiple time and length scales. Material processing involves the optimization of a number of process parameters. In addition, most materials are multicomponent systems. Processing parameters and chemistry may affect the final structure (grain size, volume fraction) of the desired phase.

As a result, uncertainty can spread over length and time scales. Uncertainty in the input of a mathematical model will affect the output of that model. In multiscale materials modeling, output of any model can be input for another model (Figure 2.1). This way, uncertainty propagates through various length and time scales, from one model to another, and affect the desired properties of the targeted materials. Bayesian approaches can be helpful in addressing uncertainty propagation.

3.2.1.2 Uncertainty Mitigation

This stage focuses on attempts to reduce the impact of uncertainty on the design of materials. A variety of uncertainties can be minimized by:

- Statistical analysis
- Surrogate modelling: Development of predictive models based on AI-based algorithms. A number of supervised machine learning approaches may be useful.
- Multidisciplinary design and optimization (MDO): Evolutionary or genetic algorithms can prove to be helpful.

3.2.1.3 Uncertainty Management

This is one of the most important stages in handling uncertainty. Statistical approaches, and AI-based algorithms can be helpful in reducing/minimizing effect of various uncertainties. But still, while dealing with real-world problems, one has to live with a certain level of uncertainty. Thus, a user needs to define an acceptable level of uncertainty for their problem. Multi-criterion decision making (MCDM) is one such approach that can be helpful in dealing with uncertainty management.

3.2.1.4   Case Studies on Uncertainties

In this book, we have included case studies dealing with the various forms of uncertainty outlined above.

- **Uncertainty quantification and propagation:** We have presented a case study using "PyCalphad" software.
- **Uncertainty mitigation and management:** All of the case studies will cover this topic.

3.2.2   COLLECTING DATA FROM VARIOUS SOURCES

For case studies considered in this book, data was acquired from various sources:

- **Experiments:** Hard magnetic AlNiCo alloys, nickel-based superalloys, soft magnetic FINEMET-type alloys, nanoindentation test, thermomechanical test.
- **Industrial furnaces:** Data was acquired from three (3) different iron-making blast furnace and one (1) steelmaking LD furnace.
- **Industry:** Data from a rolling mill from an integrated steel plant.
- **Simulations:** Sources include CALPHAD-based calculations using software like FACTSAGE, Thermo-Calc, JMATPro, phase-field simulations, DFT calculations using VASP.
- **Imaging data from experiments:** Atomic force microscopy (AFM) image.
- **Online repositories and literature:** Repositories like Github, Materials Project, VASP wiki, CITRINE informatics, and journal articles and book chapters.

3.2.3   PRE-PROCESSING/CLEANING DATA FROM VARIOUS SOURCES

While collecting data from various resources, a user is exposed to various forms of uncertainties associated with the data set. One can trust the data download from a well-maintained repository, but one must still understand how the data were collected.

- **Experiments:** Experimental equipment needs to be calibrated regularly. A user must look at the experimental reading and refer to standards prior to using this data for further analysis.
- **Industrial furnaces:** Data was acquired from three (3) different iron-making blast furnace and one (1) steelmaking LD furnace. These data are collected from harsh environments, where sometimes sensors can give faulty readings. These data sets contain lots of missing values and will have several outliers. Thus, these data sets need to be thoroughly analyzed prior to further processing.
- **Simulations:** In this book, we have included several case studies on data obtained from simulations. Sources include CALPHAD-based calculations using software like FACTSAGE, Thermo-Calc, JMATPro, phase-field simulations, DFT calculations using VASP.

- **Imaging data from experiments:** Advanced diagnostic tools like scanning electron microscope (SEM), atom probe tomography (APT), atomic force microscope (AFM), etc. generate lots of data in the form of images. These data can be efficiently utilized after preprocessing. In this work, we have included a case study on the prediction of AFM image.

3.2.4 SCALING DATA

Scaling data is an important step that can help significantly improve the predictive accuracy of AI-based models.

3.2.5 STATISTICS AND VISUALIZATION

In this book, we will provide computer codes that will be helpful in visualization of raw data set and perform various statistical calculations on the data set.

3.2.6 PREPARATION/SELECTION OF DATA FOR DEVELOPMENT OF SURROGATE MODELS THROUGH MACHINE LEARNING:= (BOX AND DRAPER 1987)

Both supervised and unsupervised machine learning algorithms have been successfully employed in the materials science domain. Supervised learning is usually associated with multi-objective optimization (Mueller et al. 2016).

A few basic terms associated with data-driven models can be listed as follows:

Training data: In supervised learning, this training data consists of input (variables) and output (objectives). Through supervised machine learning approaches, surrogate models are developed for the objective as a function of input variables, while in unsupervised learning, the entire data set is presented to the system. Here, the purpose is to find patterns in the data set.

Testing data set: This is to be used for testing the accuracy of the predictive model. In supervised learning, while training, the model is not exposed to this set of data. Hence, this data set can be used to test the accuracy of the model. If the expert is satisfied with a certain level of accuracy, then they can use these models to test even new inputs that the expert thinks can yield better results. In unsupervised learning, this data can be utilized in a different way. One can check for new patterns within the data set. Thereafter, see that it matches with the previous observations. This testing data set can then be merged with the previous data set. In supervised learning, one can use this merged data set to develop new models, while in unsupervised learning, one can use this data set to discover new patterns and observe the shift in behavior of the system.

3.3 DESIGNING EXPERIMENTS OR PREDICTING EXPERIMENTAL PARAMETERS THROUGH DATA MINING

Concept of design of experiments (DoE) was introduced by Sir RA Fisher in 1920, where the purpose is to minimize the number of experiments (mode FRONTIER). It is basically a methodology which helps in extracting valuable information from experimental data. This eliminates redundant observations and reduces the time and resources required to conduct experiments.

Applications:

1. Variable screening: To quantify the contribution index of a variable on an output.
2. Generate support points for response surface development.
3. To be used for generating an initial set of population or potential solutions while performing optimization using evolutionary/genetic algorithms.

Depending on the application, a user must select a different design of experiments approach.

3.3.1   RANDOM SEQUENCE GENERATORS: (MODE FRONTIER)

Choosing a random sequence generator among several popular approaches is again based on user preference. Statisticians and data scientists prefer a certain approach and stick to it. For our work, we mostly used Sobol's algorithm, but, in a few cases, adaptive space filler performed better when compared to other algorithms. The random sequence (RNDDOE) in modeFRONTIER distributes points uniformly in design space. This is based on the mathematical theory of generating random numbers.

3.3.2   SOBOL'S ALGORITHM (SOBOL 1967)

Sobol's algorithm is a deterministic algorithm that mimics the behavior of the Random Sequence. For all the algorithms, the final aim is to generate points that are uniformly distributed in the design space. This distribution is prone to clustering. Sobol's algorithm generates sequence in a way that clustering effects are minimized. The algorithm generates solutions that avoid each other, thus reducing the clustering and uniformly filling the design space.

3.3.3   ADAPTIVE SPACE FILLER (MODE FRONTIER)

It is useful for generating a uniform distribution of points in the input space. Incremental space filler (ISF) is an augmentation algorithm that takes into account existing points within the database (previously generated designs). Response surface generation requires appropriate distribution of support points in the design space. Through this algorithm, new points are added in a manner that the design space is uniformly filled. In this algorithm, new data points are added in a sequential manner that checks upon distance from existing points. New points will be assigned to a position in the design space where the distance is maximized with respect to the existing points. Zone filling is another feature through which uniform sampling can be achieved. A user can define a design and a radius at the initiation. This design will act as the center of the sphere. New points will be chosen in balls around this marked design. Algorithm can choose a center automatically if it is not defined.

3.3.4 IMPORTANCE OF SUPPORT POINTS IN DEVELOPING ACCURATE PREDICTIVE MODELS (BOX AND DRAPER 1987)

The design of experiments approach is useful for getting information about the problem and about the design space. They may serve as a starting point for a further optimization process, or as a database for response surface training, or to verify the sensitivity of a candidate solution. Strength assessment refers to the assessment of the effects of random variability of certain parameters on responses. The calculation of robustness is extremely important in all areas, particularly for non-linear problems. For non-linear problems, optimized solutions may be sensitive to small parameter variations. Robustness can be ascertained by applying a systematic disturbance analysis based on randomly generated values for the variables.

3.4 SURROGATE MODELS/META-MODELS

3.4.1 SELECTION OF MACHINE LEARNING ALGORITHMS BASED ON NATURE OF DATA

As mentioned throughout this book, a user needs to decide upon the AI/machine learning algorithm for developing surrogate models which can be used for prediction in future. In this book, we have included a case study on choosing a machine learning algorithm for a given set of data. This chapter will be quite helpful in understanding the correlation between data, and machine learning algorithm.

3.4.2 PERFORMANCE MEASUREMENTS OF A METAMODEL

Performance of each meta-modeling technique can be measured from the following aspects:

- Accuracy – the capability of predicting the system response over the design space.
- Robustness – the ability to achieve good precision for different types of problems and different sample sizes.
- Efficacy – computational effort required to build the meta-model and predict the response for a set of new points per meta-model.
- Transparency – ability to explicitly illustrate the relationship of input variables to responses.
- Design simplicity – easiness of implementation. Simple methods should require minimal user input and be easily tailored to each problem.

In order to provide a more comprehensive picture of the accuracy of the metamodel, three different parameters were used. Formulation for R-square, relative average absolute error (RAAE), and relative maximum absolute error (RMAE) have been provided in equations 3.1, 3.2, and 3.3, respectively.

3.4.2.1   R Square (R^2)

$$R^2 = 1 - \frac{\sum_{i=1}^{n}(y_i - \hat{y}_i)^2}{\sum_{i=1}^{n}(y_i - \bar{y}_i)^2} = 1 - \frac{MSE}{Variance} \tag{3.1}$$

where $\hat{y}_i$ is the corresponding predicted value for the observed value y and $\bar{y}_i$ is the mean of the observed values. Although the mean square error (MSE) represents the discrepancy between the meta-model and the actual simulation model, the variance indicates how uneven the problem is. The higher the R^2 value, the better the meta-model.

3.4.2.2   Relative Average Absolute Error (RAAE)

$$RAAE = \frac{\sum_{i=1}^{n}|y_i - \hat{y}_i|}{n * STD} \tag{3.2}$$

where STD refers to standard deviation. The smaller the RAAE value, the more precise the meta-model.

3.4.2.3   Relative Maximum Absolute Error (RMAE)

$$RMAE = \frac{max(|y_1 - \hat{y}_1|, \ |y_2 - \hat{y}_2|, \ ...., |y_n - \hat{y}_n|)}{STD} \tag{3.3}$$

A large RMAE indicates a significant error in a region of the computing space, although the overall precision indicated by R^2 and RAAE may be very good. Therefore, a small RMAE is preferred. However, because this measure cannot demonstrate overall performance in the design space, it is not as significant as R^2 and RAAE.

3.4.2.4   AIC (Akaike Information Criterion): (mode FRONTIER)

The AIC is used to assess the relative quality of a statistical model. It provides a relative estimation of the information lost during the generation of the model representing the actual data. It does not provide any absolute information on the extent to which a model is consistent with the data. AIC is only meaningful in comparison with other candidate models trained using the same algorithm. The best model is that which has the lowest AIC.

3.4.3   STATISTICAL APPROACHES: (MODE FRONTIER)

There are various statistical methods for the development of mathematical models. Only the Levenberg-Marquardt algorithm is covered in this book. Levenberg-Marquardt is an alternative to the Gauss-Newton method of finding the minimum of a function $F(x)$. Function $F(x)$ is a sum of squares of non-linear functions. It can be expressed as shown in equation 3.4.

$$F(x) = \frac{1}{2} \sum_{i=1}^{m} [f_i(x)]^2 \tag{3.4}$$

For a Jacobian $J_i(x)$ for a given function $f_i(x)$, the Levenberg-Marquardt method searches in the direction given by the solution p to equation 3.5.

$$(J_k^T J_k + \lambda_k I)p_k = J_k^T f_k \tag{3.5}$$

where λ_k are non-negative scalars and I is the identity matrix. In this method for some scalar Δ related to λ_k, the vector p_k is the solution of the constrained sub-problem of minimizing equation 3.6 subject to condition 3.7.

$$\frac{\|J_k p + f_k\|}{2} \tag{3.6}$$

$$\frac{\|p\|}{2} \leq \Delta \tag{3.7}$$

3.5 MACHINE LEARNING: SUPERVISED (BOX AND DRAPER 1987, MUELLER ET AL. 2016)

In this part, we discuss various methods of developing data-driven models. These models have also been referred as surrogate models or response surface models (RSM) or meta-models. Surrogate models are basically data-driven models that require an initial set of experimental data to construct the model. Hence, RSM can be defined as a statistical measure that takes into account the quantitative data from experiments to determine the behavior of the system and can be utilized to solve multi-variant equations. The RSM/surrogate models can be utilized for various applications such as:

1. Identify the factors (or system variables) that will meet a desired set of specifications in order to understand the behavior of the system under review.
2. Design Optimization: Identify various combinations of factors (or variables) that will produce a desired response area and estimate the response near the optimum.
3. Sensitivity Analysis: Determines the effect of variation of factors (or variables) on a specific response above the area of interest.
4. Uncertainty Analysis: Identify and analyze any specific region of interest responses for various combinations of factors (or variables) that have not been tested during model development.

3.5.1 ARTIFICIAL NEURAL NETWORKS (ANNS) (BASHEER AND HAJMEER 2000)

An artificial neural network (ANN) is an attempt by researchers to mimic the functional complexity of the nervous system of a human brain, with ANN being its simplest representation. A simple depiction of ANN is composed of an input layer, a hidden layer, and an output. The hidden layer processes the information provided by the input node (x) and weights associated (w) with the connection between the input node and the node in the hidden layer. This information is transmitted to output via a transfer feature. The transfer function for the final output is normally an hyperbolic tangent function. Each node in the hidden layer is associated with a bias value (w_0).

Equations 3.8 and 3.9 show the output of a single neuron g and $y(x)$. An ANN can outperform statistical methods like linear regression. An ANN is able to fit highly non-linear functions that cannot be fit by other conventional techniques.

$$g = \sum_{j=1}^{n} w_j x_j + w_0 = W^T X \tag{3.8}$$

$$y(x) = sigmoid(g) = \frac{1}{1 + exp^{W^T X}} \tag{3.9}$$

The ANN used in modeFRONTIER is based on a classical feed-forward ANN with a single hidden layer. The ANN networks are trained by the Levenberg-Marquardt algorithm. The number of nodes in the hidden layer can be set by the optimizer and/ or the user (modeFRONTIER 2021). An in-house developed version, evolutionary neural network (EvoNN), was also used for generating a portion of the results presented in one of the case studies (Jha et al. 2014).

Advantages:

1. It is used to fit complex functions, which is hard to fit through PR, RBF, etc.
2. It is flexible and sturdy and can be applied to a variety of domains/issues.

Limitations:

1. It requires a large training set.
2. For highly non-linear and high-dimensional problems, the number of nodes in the hidden layer, or number of hidden layers will increase. It will result in a highly complex model and will increase the computational cost.

3.5.2 Deep Learning (Lecun et al. 2015)

Deep learning is based on a multilayered artificial neural network. It is trained with stochastic gradient descent using back-propagation. The network can contain a large number of hidden layers consisting of neurons and other functions like adaptive learning rate, rate annealing, momentum training, dropout, and L1 or L2 regularization. This significantly improves its predictive accuracy.

We have developed our own code in the Python programming language using TensorFlow (Abadi et al. 2016) and Keras (Chollet 2015) libraries. Additionally, we also used the deep learning algorithm in two different software, Rapidminer (Rapidminer 2021) and Abacus AI (Abacus AI 2021).

3.5.2.1   Tensorflow and Keras

In this book, we used our own code developed in Python programming language. Deep learning artificial neural network (DLANN) used in this book has four hidden layers. For a given DLANN network, the first hidden layer consists of 50 neurons, and the remaining three layers consist of 100 neurons each. Figure 3.1 shows the representation of a DLANN model of this architecture. The same architecture was viewed in a visualization platform, Tensorboard. Figure 3.2 shows the representation of a DLANN model in the Tensorboard platform. DLANN architecture is varied. The initial layer has 50, 60, 70, 80, 90, or 100 neurons, while each of the other three hidden layers consist of 100, 120, 140, 160, 180, or 200 neurons, respectively.

This data set was divided randomly into a training set for training the DLANN model, and a testing set for its validation. The testing set consisted of 33% of all the available data mainly to avoid overfitting. Rectified Linear Unit (ReLU) was chosen

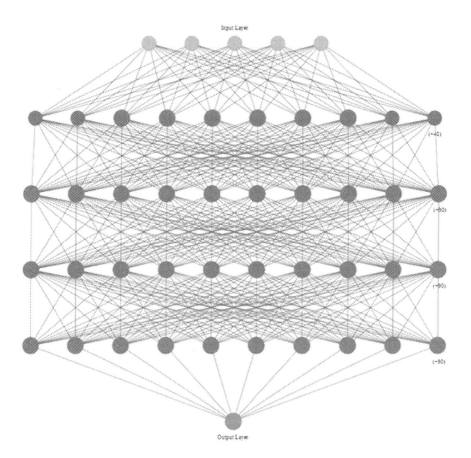

FIGURE 3.1   Deep learning neural network developed through TensorFlow/Keras.

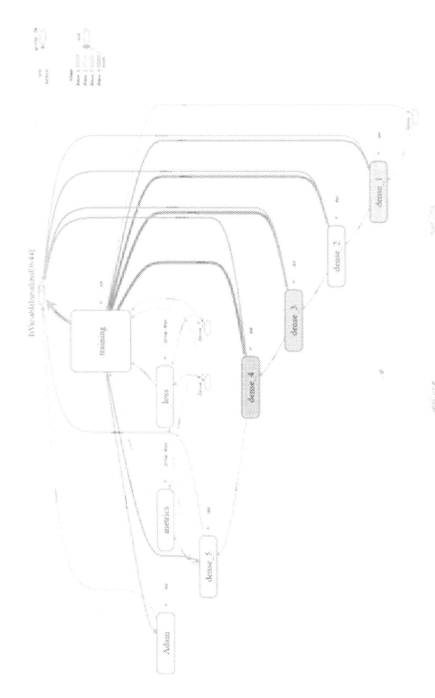

FIGURE 3.2 Tensorboard representation of deep learning neural network developed through Tensorflow/Keras.

as activation function, while for the "Adam" was chosen as an optimizer. Adam stands for Adaptive Moment Estimation, stopping criterion for training was number of epochs and it was fixed at 100 for all of these cases. Mean squared error (MSE) and mean absolute error (MAE) over the validation set was one of the performance metrics considered for choosing a model for various stable and metastable phase. Tensorboard was used for visualization and determining/choosing the model that can be used for predictive tools. Apart from MAE/MSE error metrics and visualization via Tensorboard (Tensorboard 2021), the most important criterion for choosing a model was physical metallurgy of metallic alloys. We are dealing with a very complex system of alloy design. The data set contains a lot of missing data for several metastable phases. Priority was given to the physical metallurgy of metallic alloys (aluminum, titanium). We have used the concepts of statistics and artificial intelligence as guiding tools, while avoiding being overly reliant on these tools.

3.5.3 Genetic Programming

Genetic programming (GP) is a methodology based on evolutionary algorithms inspired by biology to find computer programs that perform a user-defined task (Collet 2007, Giri et al. 2013, Poli et al. 2008). It is based on the principles of evolutionary algorithms (EAs), except that, here, every individual is a computer program with his own specialized and recombinant mechanism (crossover) and mutation. It is a machine learning technique used to optimize a computer program population based on a fitness environment determined by a program's ability to perform a given calculation task. A computer program is depicted like a tree. A tree can be considered a structure with internal and external nodes. Internal nodes are occupied by functions, while external nodes are terminals. A set of terminals consists of constants and variables, while a set of functions consists of mathematics operators and basic functions. These tree-type computer programs are bred through iterations or generations. In each iteration, it is subjected to genetic operators. Evolutionary design is a version of genetic programming that is used in modeFRONTIER (modeFRONTIER 2015). An in-house developed version, Bi-objective genetic programming (BioGP) was also used for generating a portion of the results presented in one of the case studies.

Advantages:

1. It is used to fit complex functions, which is hard to fit through PR, RBF, etc.
2. It relieves our load of algorithms development that will lead to optimized functions.
3. Pairing with EA improves its performance to accommodate highly non-linear functions.
4. It is flexible and robust and can be applied to a range of areas/issues.

Limitations:

1. It requires a large training and a testing set. In materials design, one has to deal with lots of scarce data sets.

2. After a certain iteration, the tree grows and becomes thicker without significantly improving its performance. This phenomenon is generally known as bloating. This will add to the computing cost.

3.5.4   K-Nearest Neighbor

This is one of the most well-known algorithms for partitioning and clustering, which is also used for regression. In the K-Nearest algorithm, interpolation is based on the K-nearest designs in the vicinity of a candidate points. It only takes into account a user-specified number of nearest neighbor points, K, while performing computation. It works on the basis of using the weighted sum of the K nearest points, as expressed in equation 3.10.

$$S(x) = \sum_{i=1}^{N} \omega_i(x) f(x_i)$$

(3.10)

$$\omega_i(x) = \frac{Px - x_i P}{\sum_{i=1}^{K} Px - x_i P^{-p}}$$

(3.11)

It can be seen from equation 3.11, that the weights are obtained by the normalized inverse power p of the distances. In this study, K was kept constant as 7 and 11 and p as two for all of the test cases (modeFRONTIER 2021).

3.5.5   Kriging Model (modeFRONTIER 2021)

Kriging models are widely used for estimation of inconsistent data. The proxy function is a combination of a global and local proxy model. It is the combination of a polynomial function with its departure. It can be represented in equation 3.12:

$$\hat{y} = \sum_{j=1}^{k} \beta_j f_j(x) + Z(x)$$

(3.12)

$$Cov[Z(x_i), Z(x_j)] = \sigma^2 R(x_i, x_j)$$

(3.13)

where Z(x) is assumed to be a realization of a stochastic process with mean zero and spatial correlation function given by equation 3.1. The Gaussian correlation function is the most popular and is widely used, whereas other correlation functions may also be used. Usually $f_j(x)$ is a constant term.

Advantages:

1. Given that the KG function includes both a trend function and its deviation, it is very useful for predicting temporal and spatially correlated data.
2. It is fairly flexible due to the availability of a wide range of correlating functions.

3. It provides a stepwise algorithm basis for determining critical system variables and the same data can be used to construct the predictive model.

Limitations:

1. Building the model takes time and increases the computational cost.
2. For highly non-linear and large problems, the calculation time is high if the initial data set is large.
3. It is possible for the correlation matrix to become singular if the sampling points are placed close to one another or are generated from a particular plane.

3.5.6 GAUSSIAN PROCESSES (modeFRONTIER 2021, SCHULZ ET AL. 2018, TANCRET 2012)

Gaussian processes are being used along with ANNs for developing regression models. Bayesian approaches have been successfully implemented to develop regression models based on ANN, as well as Gaussian prediction models. Gaussian processes are based on the Bayes method of probability distribution. In other words, it may be regarded as a generalized model of Gaussian probability distribution. However, they are best suited for non-polynomial responses. In the text below, we present a general description of the algorithm of a process where the average is supposed to be zero. For a given data set corresponding to a non-linear function $y(x)$, input vectors X_N and output vectors t_N are denoted in equation 3.14.

$$X_N = [x_1, x_2, x_3, x_N]; t_N = [t_1, t_2, t_3, t_N] \qquad (3.14)$$

Posterior probability distribution of $y(x)$ can be expressed as denoted in equation 3.15.

$$P(y(x)t_N, X_N) = \frac{P(t_N y(x), X_N) * P(y(X))}{P(t_N \quad X_N)} \qquad (3.15)$$

In order to predict the future values of t, it is important to know the assumed prior distribution of the function, $P(y(X))$, and the assumed noise model, whereas the parametrization of the function $y(X W)$ is irrelevant (for a parameter W). The basic idea is to put the prior on the area of the function without parameterizing. In this case, the most straightforward prior type will be called the Gaussian process. Gaussian processes are specified by mean and covariance works in the same manner as the Gaussian distribution has mean and covariance matrix. Here, the mean is a function of x, and the covariance can be estimated by evaluating the function $y(x)$ at point x and x'. Thus, the expected covariance can be denoted as $C(x, x')$.

3.5.7 HYBRID META-MODELS (modeFRONTIER 2021)

The HYBRID algorithm in modeFRONTIER combines the fittest polynomial RBF and the Kriging formulation into one hybrid model. The fittest polynomial RBF is used as a transfer function in the DACE (Design and Analysis of Computer Experiments) Kriging formulation. This form of Kriging formulates the correlation function, as shown in equation 3.16 (Dulikravich and Colaço 2015).

$$\varphi(|x_i - x_j|) = \sum_{i=1}^{n} \Theta_k |x_i - x_j|^{p_k} \tag{3.16}$$

Here, both Θ_k and P_k have to be optimized.

3.5.8 RADIAL BASIS FUNCTIONS (RBF) (MODEFRONTIER 2021)

It is a real valued function, whose value depends on the distance from the origin or any other center in the design space. The distance is usually the Euclidean distance, whereas a few other metrics can also be used. Radial basis functions (RBF) are one of the most popular mesh-free kernel approximation techniques. Originally, RBFs were developed for dispersed multivariate data and function interpolation. Later, it was found that RBFs were capable of constructing an interpolation scheme with favorable properties. The benefits include high efficiency, good quality, and the ability to process dispersed data, in particular for higher-dimensional problems.

RBF may be classified into two main classes: globally supported and compactly supported:

- Gobally supported includes multiquadrics (MQs), inverse multiquadrics, thin plate splines, Gaussians, etc.
  MQ: $\sqrt{(x - x_j)^2 + c_j^2}$, where c_j is a shape parameter.
- Compactly supported: Wendland family $(1 - r)^n_+ + p(r)$, where $p(r)$ is a polynomial and $(1 - r)^n_+$ is 0 for r greater than the support.

For a function of L variables x_i, $I = 1, ..., L$. A globally supported RBF model can be formulated as

$$s(X) = f(X) = \sum_{j=1}^{N} \alpha_j \varnothing(|X - X_j|) + \sum_{k=1}^{M} \sum_{i=1}^{L} \beta_{i,k} p_k(x_i) + \beta_0 \tag{3.17}$$

$f(\mathbf{x})$ is known for set of $\mathbf{x}$ values, where $\mathbf{x} = \{x_1, ..., x_i, ..., x_L\}$

For a given basis of polynomials, $p_k(x_i)$ is one of the M terms. There are two unknown values, α_j and $\beta_{i,k}$, which are estimated by solving a system of N linear equations. Constraints are as follows:

$$\sum_{j=1}^{N} \alpha_j p_k(x_i) = 0$$
$$\vdots \tag{3.18}$$
$$\sum_{j=1}^{N} \alpha_j p_k(x_L) = 0$$

TABLE 3.1

RBF Approximation Used in this Work

Type of approximation	$\varphi(|x_i - x_j|)$
Multiquadrics (MQs)	$\sqrt{(x_i - x_j)^2 + C_j^2}$
Inverse Multiquadrics (IMQs)	$\left\| \sqrt{(x_i - x_j)^2 + C_j^2}\, \right\|^{-1}$
Gaussian	$exp\left[-C_j^2 (x_i - x_j)^2\right]$

$$\sum_{j=1}^{N} \alpha_j = 0 \qquad\qquad (3.19)$$

Table 3.1 summarizes some commonly used RBFs. In this table, an abbreviated name is provided for each approximation, which may be used later in the benchmark analysis.

Advantages:

1. It is suitable for a wide variety of applications. It provides superior performance for high-order non-linear problems tested for large-scale/small-scale data.
2. It has been successfully implemented for continuous and discrete response functions.
3. Radial basis function network can be assumed as an ANN with a single hidden layer. RBFs have been used as kernels for support vector mechanics.

Limitation:

1. Its performance is relatively poor outside the variable bounds, especially in the absence of a polynomial function.

3.5.9 POLYNOMIAL REGRESSION MODEL (PR) (MODEFRONTIER 2021)

PR models are widely used by researchers to design complicated engineering problems. Objectives are mathematically (statistically) related with the design variables. For a problem of one variable, PR models are curve fitting. On the basis of problem, one can develop PR models of first and second order. It can be formulated as shown in equation 3.20:

$$y = f'(x)\beta + \epsilon \qquad\qquad (3.20)$$

Here, y is the response, $x = (x_1, x_2, \dots x_k)'$, $f(x)$ is a vector function of p elements that consists of powers and cross products of powers of $x_1, x_2, \dots x_k$, up to a certain degree denoted by $d(\geq 1)$, β is a vector of p unknown constant coefficients/parameters, and ϵ is a random experimental error assumed to have zero mean.

PR models of first degree ($d = 1$) and second degree ($d = 2$) are shown by equations 3.21 and 3.22.

$$y = \beta_0 + \sum_{i=1}^{k} \beta_i x_i + \epsilon \qquad (3.21)$$

$$y = \beta_0 + \sum_{i=1}^{k} \beta_i x_i + \sum_{i<j} \sum \beta_{ij} x_i x_j + \epsilon \qquad (3.22)$$

For developing a response surface, β needs to be estimated. It is usually estimated by regression or least square techniques.

Advantages of using PR models:

1. Coefficients associated with the normalized regression models can be helpful in determining significance of system variables.
2. A user can choose between PR models of different orders based on the complexity of the problem and save on computational cost. In this book too, we have used curve fitting or AI-based supervised machine learning models based on complexity of the problem.
3. For high-dimensional problems, second-degree polynomial is to be used.
4. Convergence is fast for PR models.

Limitations:

1. PR models' performance is not as per expectation while predicting the response outside the variable bounds.
2. Higher-order polynomials can be used for highly non-linear problems. Determining all the coefficients of the PR equation can be difficult in a lot of cases.

3.6 MACHINE LEARNING: UNSUPERVISED (MODEFRONTIER 2021)

There is no training process, so these algorithms are known as unsupervised machine learning algorithm. Unsupervised machine learning algorithms are used for discovering patterns and clusters in a data set. It is often used for reducing the dimensionality of the data set. Reducing dimensionality helps in visualizing a high-dimensional data and better understanding of the physics of the problem. One must refer to the literature while interpreting findings from these approaches. Understanding the basic physics of the problem is the key for potential implementation, as there is always some loss of information while reducing dimensionality. Loss of information can be minimized by different scaling technique. In this book, we have used three classes of unsupervised machine learning algorithms in the case studies.

In the following text, we explain the theoretical aspects of these approaches. Its application will be discussed in the case studies.

3.6.1 Principal Component Analysis (PCA) (modeFRONTIER 2021)

PCA is one of the most widely used unsupervised machine-learning algorithm (Mueller et al. 2016). PCA is used for determining correlations between variables and objectives in high-dimensional data. It is capable of reducing the dimensionality of the high-dimensional data set without losing much information. An orthogonal transformation converts a set of mostly correlated variables and objectives into a set of linear uncorrelated variables. This set of uncorrelated variables is known as principal components (PCs). Each principal component is a linear combination of all the original data descriptors (variables and objectives). The first principal component (PC1) is associated with maximum variance in the data set, followed by PC2, PC3 …. PC (n-1), where n is the dimentionality of the original data set (modeFRONTIER 2021). Usually the first few components are associated with most of the variance. Other components are also important and must be taken into account, but it is possible to visualize a high-dimensional data set by choosing just PC1, PC2, and PC3 (Mueller et al. 2016). It is also used to identify patterns, clusters, or trends in the data because trends can be difficult to find in large high-dimensional data sets.

Three important terms associated with PCA analysis:

- Scree plot: A graph between eigenvalues and the component number. This is an important chart that helps determine how many components are needed to represent the complete data set. Typically, components that have eigenvalues greater than one (1) are chosen for further analysis. The scree graph usually flattens below eigenvalue 1, which means that subsequent components do not have a significant effect on the data set. Given that each successive component represents a comparatively smaller variance, the least influencing components may be ignored in the subsequent analysis.
- Eigenvalue: It is the variances associated with the principal components. Principal components analysis is conducted on the correlation matrix. The variables are standardized, so that each variable has a variance of one, and the total variance is equal to the number of variables used in the analysis. Therefore, the number of principal components extracted from a data set is one less that the sum of the number of variables and objectives in that data set. The first principal component is associated with the highest variance and highest eigenvalue. The following components will take into consideration the greatest possible portion of the remaining variance. Therefore, each successive component will represent comparatively less variance (thus less eigenvalue) than the one leading to it.
- Component Graph: After selecting the required number of components, these factors are plotted against one another, while the original variables and objectives are plotted on this abbreviated space. In modeFRONTIER, each variable and objective is represented by an arrow. The length and orientation of this arrow on the reduced dimensional space determines the contribution of that variable or objective towards a certain principal component. If an arrow is positioned along PC1 on the 0-line and is perpendicular to PC2, then this variable or objective will have maximum contribution on PC1 and minimum contribution towards PC2.

3.6.2 HIERARCHICAL CLUSTERING ANALYSIS (HCA) (MODEFRONTIER 2021)

HCA is another important clustering etchnique that finds application in materials science domain (Mueller et al. 2016). Each data point acts as an initiation site for clustering. These clusters are fused iteratively to form larger clusters and eventually merged into one large cluster. The final output is a tree structure called a dendrogram, which shows how clusters are linked. The user may specify a distance or number of clusters to display the data set in disjointed groups. This way, the user can get rid of a cluster that serves no purpose, depending on their expertise. HCA analysis was performed in modeFRONTIER (ESTECO 2021) software.

Clusters are classified by the following measures:

1. Internal similarity (ISim): It is a measure of the compactness of the k-th cluster. It must be higher.
2. External similarity (ESim): It reflects the uniqueness of the k-th cluster. It must be lower.
3. Descriptive variables: They are the most significant variables that help in identifying cluster elements that are similar to one another.
4. Discriminating variables: They are the most significant variables that help in identifying cluster elements that are dissimilar to other clusters.

3.6.3 SELF-ORGANIZING MAPS (SOM) (MODEFRONTIER 2021, JHA ET AL. 2017)

Self-organizing feature maps (SOM) are a classification technique based upon an unsupervised neural network. It was proposed by Teuvo Kohonen in the 1980s. SOM is best known for preserving the topology of the data set through competitive learning and neighborhood function. It is one of the best-known tools for visualizing high-dimensional data sets in lower dimensions. With careful training, there is minimal loss of information with respect to the topology. In a few case studies, we have reported a topological error of 0.00001. Throughout the book, we have mentioned that quantization error associated with SOM is on the higher end. Thus, one can visualize a high-dimension non-linear data set in two-dimensions. Since, topology preservation is perfect for SOM, it can be used to determining various correlations within the data set. SOMs may be regarded as a non-linear generalization of the principal component analysis. In the materials engineering domain, particularly in alloy design, there is always scarcity of data. A conventional neural network is a powerful tool, but it requires a large data set for training. SOMs have been successfully used for feature extraction especially in scarce data sets with a sample size of 40. This makes it a perfect tool for materials and metallurgical engineering applications. In alloy design, we can have a scarce data set with respect to number of candidate alloys. But, most alloys in application may consist of 10–20 elements, and for applications this alloy needs to possess multiple superior properties; thus, a scarce data set with lots of variables and objectives. SOM is a perfect tool for this kind of problem. In many cases, it is extremely complex to determine correlations between several variables and objectives. Due to SOM's topology preservation, various correlations can be visualized. Even if quantization error is high, one can just use SOM for determining patterns and for a predictive model they can refer to a supervised machine learning approach. In this book, SOM analysis was performed in modeFRONTIER.

SOM analysis uses the following steps:

- Scaling: Scaling is an extremely important parameter. By choosing "logistic" scaling, errors were significantly reduced for a number of problems in our work. There are other scaling techniques in modeFRONTIER that can also be used. In logistic scaling, a transformation is linear around the average value and smoothly non-linear at the variable bounds.
- Learning cycle: There are two options: sequential and batch mode. Both of them can be used. In this book, we have used the batch mode.
- Set up training parameters: Other parameters that can be adjusted in modeFRONTIER:
  a. X-dimension and Y-dimension: It is an integer positive value. It can be set by the user or the system can calculate it by default on the basis of size of data set. In the default mode, the ratio between X and Y is equal to the ratio between the first two principal eigenvalues of the data set.
  b. Map units: Equal to X units times Y units.
  c. Initialization type: Both linear and random initialization were tried in this book. In the linear mode, initial map is obtained by using a linear combination of Kohonen map dimension and the two principal eigenvalues of the data set.
  d. Random seed: Seeding is an important parameter. It is an integer number and setting the same number will repeat the sequence. In AI-based algorithms, there is always an emphasis on promoting randomness in the design space. This can be achieved by setting a random seed value to 0. For a value of 0, the system refers at the current time, which is always different. Thus, same sequence will not be repated during SOM map generation.

There exist a few more parameters that can be fine-tuned for minimizing error or left at default. Parameters include initial rough radius, final rough radius, rough phase radius, initial fine-tuning radius, final fine-tuning radius, and fine-tuning phase length.

SOM maps were used for the following reasons:

- Identify correlations among various variables and objectives that may be supported by literature.
- Identify different clusters and study these individual units rather than analyzing the complete data set.

Applications in this book:

- Predict chemical composition of alloys for optimum set of properties.

- Determined correlations between process parameters in an industrial furnace and rolling mill.
- Demonstrated the potential of using it as an additional screening tool.

3.7 MULTI-OBJECTIVE OPTIMIZATION

Most of the practical, real-world problems involve more than one objective. These objectives are, for the most part, conflicting and must be achieved to solve a particular problem. A multi-objective optimization problem may involve a number of objective functions. These objectives are to be optimized, i.e. maximized or minimized simultaneously or bound under constraints. A user can place additional constraints on the variables or objectives. Mathematically, a multi-objective problem can be formulated as shown in equations 3.23–3.26 (Deb 2001).

$$Maximize, \quad Minimize, \quad F_m(X), m = 1, 2, \dots . M \tag{3.23}$$

$$g_i(X) \geq 0, j = 1, 2, \dots J; \tag{3.24}$$

$$h_k(X) = 0, k = 1, 2, \dots K; \tag{3.25}$$

$$x_i^L \leq x_i \leq x_i^U, i = 1, 2, \dots . n. \tag{3.26}$$

The solution X is a vector of n decision vectors defined as $X = (x1, x2, x3, â.$ Variable x_i is bounded by a lower bound, x_i^L, and an upper bound, x_i^U. Regarding constraints, $g_i(X)$ is the inequality constraint and $h_k(X)$ is the equality constraint. Solutions included within the constrained variable space are known as feasible solutions. The rest of the solutions are classified as infeasible solutions.

3.7.1 PARETO-OPTIMALITY

Multi-objective optimization algorithms operate on the concept of dominance. Any two solutions are compared on the basis of their function values to determine which solution dominates over the other (Deb 2001).

A random solution, $x^{(1)}$, can dominate another solution, $x^{(2)}$, if the solution $x^{(1)}$ is not worse than $x^{(2)}$ in all objectives, i.e. $f_j(x^{(1)}) \geq f_j(x^{(2)})$ for all $j = 1, 2â.$ Additionally, solution $x^{(1)}$ must be strictly better than $x^{(2)}$ on at least one objective, i.e. $f_j(x^{(1)}) > f_j(x^{(2)})$ for at least one $j \varepsilon 1, 2â.$

3.7.1.1 Pareto-Optimal Set

After optimization, a user will end up with a pool of feasible solutions P. All of these solutions are compared among them for dominance. Solutions are compared and the end result is a non-dominated set of solutions P'. Where P' is the set of solutions that cannot be dominated by any member of the parent pool P. This non-dominated subset, P', is known as the Pareto-optimal set, Pareto-set, or Pareto Front.

For solutions included in the Pareto-optimal set, improvement on one objective will lead to deterioration in at least one of the other objectives. Thus, in multi-obejective optimization, a user will not get a unique solution to the problem; rather, they will get a set of solutions.

Local Pareto-Optimal Set: In the variable space, several solutions can form a cluster. An optimization algorithm may get stuck in that region and provide with a set of solutions that are non-dominated with respect to each other in that particular cluster which is just a subset of feasible search space. This set of locally non-dominated solutions is referred as a local Pareto-optimal set.

Global Pareto-Optimal Set: It includes a set of non-dominated solutions that were compared with all the feasible solutions in the design space.

3.7.1.2   Evolutionary Algorithms (EA) for Multi-Objective Optimization

Conventional optimization methods usually convert the multi-objective optimization problem into a one-objective optimization problem, thus insisting on a particular set of Pareto-optimum solutions for one objective at a time. When this method is used to find more than one solution, it must be applied repeatedly. Furthermore, solutions can be found in close proximity to the pre-existing solution (which we obtained in the previous simulation) in each simulation cycle, thus affecting diversity. As a result, it is difficult to ensure that the solutions we have can be considered as the global Pareto-optimal set, even after multiple cycles.

The disadvantages of traditional optimization methods can be taken into account in evolutionary algorithms by the following:

1. The population-based approach helps to find a variety of solutions.
2. Strategies for niche preservation help maintain diversity.

In contrast to conventional optimization techniques, evolutionary algorithms use a population-based approach, making them capable of simultaneously evolving multiple solutions that approach the non-dominant Pareto in a few passes. The genetic operators on this population are recombination (crossover) and mutation. These operators alters the structure of the solutions to generate new pool of solutions which are diverse in nature. It also helps in verifying that the set of newly evolved solutions does not converge prematurely towards a set of local Pareto-optimal solutions. These EA capabilities enable them to find a diversified set of solutions for challenging problems with discontinuous and multi-modal solution spaces. Moreover, most multi-objective EAs do not require the user to have a prior knowledge of the physical parameters and equations that govern the problem they face. Their characteristics make EA one of the most popular heuristic approaches to dealing with multi-objective design and optimization issues.

3.8   EVOLUTIONARY ALGORITHMS (EAS)/GENETIC ALGORITHMS (GAS)

Evolutionary algorithms (EAs) or genetic algorithm (GAs) are heuristic search algorithms that essentially draw on the Darwinian theory of evolution based on the fittest survival. In nature, every generation has individuals that are dominant over others in some respect and over their previous generations. But, nature does not discard individuals that do not belong to this elite class. The same concept is applied in EA/GA, while evolving a solution. Superior solutions are usually preferred in the selection process to improve their efficiency in the next generation, whereas the other solutions are not completely discarded and some of these are preserved to maintain diversity (Deb 2001).

Genetic algorithms are a class of evolutionary algorithms (EAs) that generate solutions for optimization issues using techniques inspired by natural evolution, such as selection, crossover, and mutation. A population of potential solutions is bred through a certain number of generations (iterations) on the basis of their fitness value. Genetic operators like selection, crossover, and mutation operate on the whole population in each generation. After a preset number of iterations, the individuals left are the best possible solutions of the problem in hand. These solutions evolved in subsequent generations so they can be considered the best. But still, one may further screen these solutions to obtain the Pareto-optimal set. In practical problems, it is necessary as one cannot practically use all the recommended solutions as it can be in thousands or even millions.

Basic terminology: Key aspects of the EA include:

1. Population: The first step when using GA is initializing a population. The population consists of all individuals in the predefined search area. All candidates constitute the possible solution to the optimization problem. Candidate solutions must be either binary or real-coded.

2. Fitness: Each individual is a potential solution of the problem and each of them is assigned a scalar value also referred to as fitness. This is an indication of its importance in the population.

3. Selection: The selection operator selects individuals in the population according to their reproductive fitness. Higher fitness translates into a higher probability of being selected.

4. Crossover or Recombination: At any instant in ageneration, two individual solutions (parent) can be selected for reproduction for creating two offspring (children) solutions. The evolved offsprings will have a different structure and a different fitness value. Generally, crossover probability is kept high (≤ 0.8).

5. Mutation: In mutation, an individual is randomly selected and its structure is altered by a preset amount. Mutation can be beneficial in some cases where a minor structural change is necessary to achieve the desired solution. Various types of mutation operators are in practice for binary as well as real coded individuals. Mutation probability is kept low (≤ 0.3).

6. Elitism: In elitist GA, the best solutions are simply copied into the next generation without modifying their structure. An individual evolves through generations. Genetic operators operate in each generation. Thus, there is a probality that genetic operators may completely destroy a solution evolved through generations by completely altering their structure. This can make the algorithm a kind of random search and we can fail in obtaining the desired set of Pareto-optimum solutions. To prevent such a situation, the term *elitism* is introduced.

Elitism can be introduced by mixing parents and offspring after each generation, and then sorting out the best individuals to replace the preceding parent
solution. Copies of the best individuals of a particular generation can also be
made and kept for the next generation. Genetic operators can be applied to the
rest of the population, saving the elites from being destroyed.

7. Generation: In mathematical terms, these are iterations. In the GA, a number
   of generations have typically been used as a stop criterion if no other stop
   criterion is defined.

3.8.1 Non-Dominated Sorting Genetic Algorithm (NSGA-II)

NSGA-II is an example of elite preserving GA. In NSGA-II, a strategy for elite
preservation and a mechanism for diversity preservation ensures better spread of the
solution (Deb 2001).

The various stages of the algorithm are as follows:

1. The first step consists of defining the population within the feasible research
   space.
2. This population is sorted into various levels of non-domination. Each solution
   is then assigned a fitness that is equal to its non-domination level.
3. Crowding distance is computed for every members of the population.
4. On the basis of crowding distance, individuals are selected for recombination
   (crossover) through binary tournament selection.
5. Genetic operators like recombination and mutation ensure that population is
   consistent over generations.
6. Parents and offspring population are mixed together. Therafter, non-
   dominated sorting is performed on the combined population. Population is
   sorted in different non-dominated fronts on the basis of preset dominance
   criterion, where each solution is assigned to a font that is equal to its non-
   domination level.
7. The parent population will be replaced by the solutions in the non-dominated
   fronts in an increasing order starting from the front 1. The last front which
   cannot be fully occupied is arranged according to the crowding distance of the
   solutions comprising it in a decreasing order of magnitude. The remaining
   positions are filled from this sorted list from the top. The remaining solutions
   to this front, as well as other fronts that cannot be included in the population,
   are rejected.
8. This loop is repeated by moving to stage 2 and continuing until the termi-
   nation criterion is met.

Some unique features that have been incorporated into this algorithm are discussed
in the following text.

3.8.1.1 Elitist Preserving Strategy

In each generation, parents and offsprings are mixed prior to sorting solutions for
the next generation. Due to mixing, sorting is performed on a population of twice

the initial size. Thus, the best evolved (elite) solutions of the previous generation have a good chance of finding a place on one of the fronts. As a result, the elites of the preceding generation have the opportunity to be part of the next generation and to preserve them.

3.8.1.2 Crowding Distance Selection

It is the representation of the number of individuals surrounding a particular individual on the Pareto front. The average distance between the two points lying on either side of the concerned individual along each of the objectives is taken into consideration for distance calculation. The two individuals on either side of the individual are located on opposite corners of a cuboid. A cuboid can be constructed in the objective space without including any other point apart from these three. Crowding distance assigned to an individual is the distance between the two individuals on the opposite corner of the cuboid. A greater crowd distance helps to preserve diversity and thus a better distribution of solutions is obtained.

In another approach, none of the fronts are guaranteed full representation in the new population. In this way, extra spaces are created in each non-dominated set. This space is to be filled by individuals solutions that were assigned a lower rank during sorting. It is an effective step in preserving diversity and ensuring a better distribution of the solution.

3.8.2 PREDATOR–PREY GENETIC ALGORITHM (PPGA) (PETTERSSON ET AL. 2007)

PPGA is different from a conventional GA as a concept of predator-prey is introduced, instead a dominance check for assigning fitness to a candidate solution.

Stage one is the initialization of a prey and predator population. Prey in this case is the usual GA population that can be a set of solutions that can reproduce and multiply over subsequent generations. While a target prey population is maintained, the prey population can fluctuate during the calculation. Predators are a family of external entities whose sole purpose is to prune the prey population in accordance with a fitness value. They have a fixed population, which means they cannot multiply or be killed. Predators and prey are randomly placed on the vertices of a 2D-toroidal lattice whose ends are contiguous, as shown in Figure 3.3.

Following initialization, the predator-prey approach shall proceed as follows:

1. Prey is permitted to find prey in its vicinity for breeding. A prey can only breed with another prey located in its neighborhood (Moore's neighbor in this case, Figure 3.4). If the prey has no neighbor, it is not authorized to breed. All other prey can breed with another randomly selected neighbor to produce two offspring from both parent prey. Here, parents and children are retained and parents are not replaced with children. This offspring is then randomly positioned on the lattice.

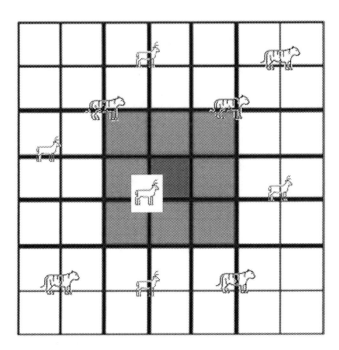

FIGURE 3.3   2-D lattice with randomly distributed predators and prey.

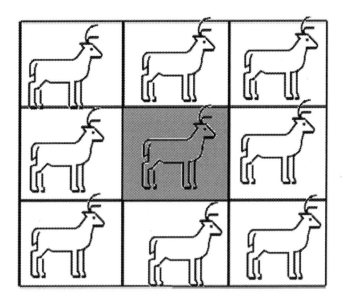

FIGURE 3.4   Mooree's neighborhood.

2. Now the predators assign fitness to the parents and offspring in its neighborhood as perequation 3.27. Predators prune the prey population based upon their fitness.

$$fitness = w \times f1 + (1 - w) \times f2 \qquad (3.27)$$

Here, w is the random value assigned to the prey by the predator and $f1$ and $f2$ are the two objective function value of the prey. Every predator has its own preference for hunting and it is determined by the w values associated with a predator.

$$No.\ of\ assigned\ Moves\ for\ Predator\ 1 =$$
$$\frac{no\ of\ current\ prey - no\ of\ prey\ refered}{Number\ of\ predator} \qquad (3.28)$$

3. Predators and prey are allowed to mix and move within the lattice from one cell to another. This movement is dictated by a random move probability. Each predator and prey has a fixed number of opportunities to find a new cell. If the cells they are attempting to occupy are preoccupied, they are left with one fewer chance out of the assigned number of chances to find a new cell. If the prey cannot find a new cell after the assigned number of chances, it will stay where it is.
4. The predator kills the prey with the lowest fitness value and takes its position. Each predator is assigned a fixed number of opportunities and can only hunt in its own neighborhood. Predators first look around the neighborhood for prey. In the presence of the prey, the predator chooses the most unfit prey and kills it. The predator then occupies the cell that was formerly occupied by this prey. A predator loses one of its assigned numbers of chances to kill the prey, if there is no prey in the neighborhood. It travels the exact same way as prey. The odds of a predator can be computed as shown in equation 3.28.
5. Excessive growth of the prey population adds to the computational complexity. Arrangements are made to remove prey after a number of generations on a regular basis based on ranking. The Prey are classified after several generations and the prey with a higher rank are eliminated from the population in order to converge the solutions towards the optimal value. This regular removal of prey also reduces the computational load.
6. The above steps are repeated until the stopping criterion is not reached.

3.8.3 Multi-Objective Particle Swarm Optimization (MOPSO)

Particle swarm optimization (PSO) belongs to the broader class of swarm intelligence techniques that are used to solve optimization problems (Kennedy and Eberhart 1995, modeFRONTIER 2021). It differs from other evolutionary algorithms, as in PSO there is no selection operator. All members of the population have the same significance and there is no specific preference for a member on any basis. This is a population-based stochastic technique that is essentially based on natural patterns observed in flocks of birds or schools of fish. In PSO, potential solutions referred to as particles, move in the search space of an optimization problem under consideration. During the initialization stage, each particle is assigned with a random initial position and an initial velocity. This algorithm also keeps track of the particle that leads the entire herd at all times. Each of the particles memorizes the position of the best solution that they found and position of the global leaders. Each particle uses its own experience and the experience of its neighbor particles to choose the manner in which it must move in the search space. At the end of each iteration, each particle updates its velocity on the basis of its own best performance so far and the global best performance of the swarm as a whole. This velocity is a weighted sum of three components: the old velocity, a velocity component that drives the particle towards the location in the search space where it previously found its best solution so far, and a velocity component that drives the particle towards the location in the search space where the neighbor particles found the best solution so far that is the global best performance of the swarm as a whole.

The velocity ($V_i^{t+\Delta t}$ and position ($x_i^{t+\Delta t}$ of the ith particle at time t are updated to time $t + \Delta t$ according to the following two equations respectively:

$$V_i^{t+\Delta t} = \omega V_i^t + R_1 \tau_1 (V_{i,IBST}^t - V_i^t) + R_2 \tau_2 (V_{i,GBST}^t - V_i^t) \qquad (3.29)$$

$$x_i^{t+\Delta t} = x_i^t + V_i^{t+\Delta t} \qquad (3.30)$$

where ω denotes user-defined inertia weights, τ terms are the used defined constants, while R terms are random numbers generated uniformly in the range [0, 1]. The term $V_{i,IBST}^t$ denotes the individual best performance of the particle so far, while $V_{i,GBST}^t$ denotes the global best performance while considering all the particles of the swarm. In equation 3.30, the first term is responsible for the inertia effect, while the second and the third term are responsible for the acceleration effects.

Multi-objective particle swarm optimization (MOPSO) (Coello Coello et al. 2002) is one of most successful and widely applied algorithms belonging to the family of swarm intelligence. Several variants of PSO algorithms exist, which are capable of solving multi-objective optimization problems for real-world problems. In this book, we have presented results using the optimization software modeFRONTIER (modeFRONTIER 2021).

3.8.4   SIMULATED ANNEALING (SA)

Annealing refers mainly to the slow cooling process of the molten material (Kirkpatrick et al. 1983). If a solid is heated to the melting point and cooled, the structural properties of the solid depend on the rate at which it is cooled. If the liquid is cooled rapidly (quenched), the crystals will have imperfections. However, if the melt is cooled down slowly enough, large crystals will be formed, which will allow the atoms to achieve a minimal energy configuration. At a temperature T, the atomic energy (E) of a substance is distributed in accordance with the Boltzmann equation, where k is the Boltzmann constant.

Simulated annealing is fundamentally a search algorithm rather than an evolutionary algorithm. It derives its inspiration from the metropolis algorithm (Metropolis et al. 1953). In the metropolis algorithm, the Boltzmann equation is used as the probability of selection for the acceptance of uphill motions in a search space. Downhill motions are also acceptable. Uphill motions are accepted only if a random number in the range [0, 1] is less than the value of the exponential term indicated in equations 3.31 and 3.32. In equation 3.32, d is the energy difference, the difference between the uphill objective function value and the function value of the base point. The value of β depends on the problem and it is to be ascertained empirically, whereas T is the temperature. One can observe that Θ decreases with an increase in the value of d increases or a decrease in value of T. Uphill movements are given a small preference in order to search the entire search area and get the exact activation energy curve.

$$P(E) = exp\left(\frac{E}{kT}\right) \tag{3.31}$$

$$\Theta = exp\frac{-d.\beta}{T} \tag{3.32}$$

3.8.4.1 Multi-Objective Simulated Annealing (MOSA)

Originally, SA was designed to use only one search agent and in some cases worked better than EA for single objective optimization. It has scarcely been used for multi-objective optimization because of its inability to find multiple solutions. The goal of multi-objective optimization is to find a well-distributed set of solutions known as Pareto front. Multi-Objective SA (MOSA) uses the concept of dominance. Annealing scheme is used for efficient search. Iterative trials helps in finding multiple solutions. The repetition of SA tests leads to its convergence towards the overall optimum with a uniform probability distribution for single-objective optimization. In the event that there are two global optimas, it can be proven that SA can find each optimum with a probability of 0.5. MOSA can find a small group of Pareto solutions within a short timeframe. After that, repeat the tests to find the additional solutions necessary to obtain the final Pareto (modeFRONTIER 2021).

3.8.4.2 Archived Multi-Objective Simulated Annealing (AMOSA)

Archived multi-objective simulated annealing (AMOSA) is a variant of MOSA, where the concept of archive has been introduced (Cyber Dyne S.r.l. Kimeme 2021). All the non-dominated solutions evolved until any particular generation is stored in the archive. So, in any generation, there is an archive of non-dominated solution, other solutions also referred as current solution, and new solutions evolved in that particular generation. The acceptance criterion for a new solution can be explained as follows. For a new solution, the algorithm checks its domination status and compares it with the current solution, and the solutions that are stored in the archive. A new solution is accepted in place of the current solution if it dominates the current solution. In AMOSA, the size of the archived set is limited. AMOSA algorithm aim is to obtain a limited number of well-distributed Pareto-optimal solutions. Thus, a new solution can be included in the archived set if it dominates the solution present in the archive set.

3.8.5   EVOLUTION STRATEGIES (ESs)

Evolution strategies (ESs) were developed by Rechenberg (1971) and subsequently amended by Schwefel (1974, modeFRONTIER 2021). This is also known as the German version of GA because of its origin. Previous versions were typically limited to two-member ESs because of the complexity of the solution evolution, making the process time consuming.

In the absence of crossover, an ES uses a selection and mutation operator quite efficiently in order to evolve a solution. This version of the mutation is called a self-adaptive mutation. A mutation is not constant over the generations, and it differs from generation to generation depending upon the perturbation needed to eveolve an optimum solution. Multi-membered ES (MMES) includes multiple members, while recombinative ES explores the benefits of crossover.

3.8.5.1 Multi-Membered ES (MMES)

It may be categorized in two ways, as shown in equations 3.33 and 3.34.

$$(\mu + \lambda) - ES \tag{3.33}$$

$$(\mu, \lambda) - ES \tag{3.34}$$

$$y^j = x^i + N(O, \sigma) \tag{3.35}$$

Here, μ is the size of the initial population and λ is the number of offspring to be generated from μ members of the initial population.

Mutated solution y^j is created from the initial population member following equation 3.35. In equation 3.35, O is the mean and σ is the standard deviation or strength of mutation. $N(O, \sigma)$ denotes normal distribution of O and σ.

The best μ members are chosen for the next generation in accordance with equations 3.33 and 3.34 after the evolution of λ. The difference is that in equation 3.33 both parents and children are mixed for choosing the best μ members, while in equation 3.34 the best μ members are chosen from the evolved set of λ offspring only. No mixing of solutions is performed in equation 3.34, and the rest of the solutions are discarded.

3.8.5.2   Recombinative ES

Recombinative ES introduces the concept of the crossover operator. It may be classified in two ways, as set out in equations 3.35 and 3.36:

$$(\mu/\rho + \lambda) - ES$$
$$(\mu/\rho, \lambda) - ES \tag{3.36}$$

The initial population includes randomly chosen ρ members. From this population, μ members are involved in a crossover. This crossover leads to the evolution of λ new offspring. All other relevant parameters have the same meaning as outlined previously.

3.8.6 DIFFERENTIAL EVOLUTION (DE)

The differential evolution (DE) algorithm was proposed by Storn and Price (1997). The usual notation used in a differential evolution is, "*DE/rand/1/bin*". "*rand*". It comes from the fact that the initial vector is *rand*omly chosen. Thereafter, "*1*" denotes the vector difference that is added to it and "*bin*" comes from the fact that the number of parameters donated by the mutant vector follows *bin*omial distribution. DE is a population-based optimizer that tries to solve the given problem by sampling the objective function at several randomly chosen points from the original set of points. In order to initialize a population of randomly generated vectors of size Np, we have to first define the upper and lower bounds of each of the parameters that is to be used. Each of the vectors is indexed with a number from 0 to $Np-1$. New points are generated as a result of perturbations due to the existing set of points, but the way in which these perturbations or deviations are introduced makes DE different from other evolutionary algorithms. In DE, perturbation is introduced as a result of scaled difference of randomly selected two existing members (vectors) of the same population. To create a new vector, DE adds this above perturbation to a randomly selected vector of the initial population. Now in the selection stage, this newly created vector competes with the vector of the same index from which it was created by the initial population. The winner is selected and this selection procedure continues until the last newly created vector has been tested. This is done in order to ensure improvement of the vectors in a subsequent generation as a result of perturbation brought about due to operation of genetic operators. This new population serves as the initial population for the next generation and the operation are repeated until convergence or any other stopping criterion is fulfilled.

3.8.6.1 Initialization of Initial Population

A vector in the initial population can be denoted as $(x_{j.i.g})$. $x_{j.i.0}$ $(g = 0)$ denotes the initial value of the jth parameter of the ith vector. It may evolve as set out in equation 3.37.

$$x_{j.i.0} = rand_j(0,\ 1)_*(b_{j.U} -\ b_{j.L}) + b_{j.L} \tag{3.37}$$

The random number generator, $rand_j(0, 1)$, generates a uniformly distributed random number in the range $[0, 1]$ where the subscript j ensures generation of a new random value for each parameter. $b_{j.U}$ and $b_{j.L}$ are the upper and lower bounds of the jth parameter involved in population initialization.

3.8.6.2 Differential Mutation

In order to create a trial vector $(v_{i,g})$ by mutation, we select two random vectors $(x_{r1,g}$ and $x_{r2,g})$ from the population of Np vectors and add its scaled difference to a randomly selected third vector $(x_{r0,g})$, as shown in equation 3.38.

$$v_{i,g} = x_{r0,g} + F. (x_{r1,g} - x_{r2,g}) \qquad (3.38)$$

Here, $F \in (0, 1+)$ is a positive real number that controls the mutation and thus the rate of evolution of the mutated population. Although there is no upper limit defined for F, in most of the cases its value is less than 1.

3.8.6.3 Crossover

Crossover is also referred as *discrete recombination*. Uniform crossover is usually done to compliment differential mutation. Dual crossover creates trial vectors from the parameter values that are copied from two different vectors. The DE crosses each vector with a mutant vector according to the condition given in equation 3.39.

$$u_{i,g} = u_{j,i,g} = \begin{cases} v_{j,i,j} & if \ (rand_j(0, 1) \le Cr \ or \ j = j_{rand}) \\ x_{j,i,g} & Otherwise \end{cases} \qquad (3.39)$$

Here, is the crossover probability and it is user defined. It controls the parameters that are copied from the mutant vector.

3.8.6.4 Selection

We are left with the combined population of the initial vector population and the vector population left after being operated by genetic operators. A new vector is selected for the next generation on the basis of its objective function value, as shown in equation 3.40. A binary tournament is played between vectors of the same index in order to ensure improvement in its performance.

$$x_{i,g+1} = \begin{cases} u_{i,g} & if \ f(u_{i,g}) \le f(x_{i,g}) \\ x_{i,g} & Otherwise \end{cases} \qquad (3.40)$$

3.8.7 STRENGTH PARETO EVOLUTIONARY ALGORITHM 2 (SPEA 2)

Strength Pareto Evolutionary Algorithm 2 (SPEA) was introduced by Zitzler, Laumanns, and Thiele (Zitzler et al. 2001). SPEA starts with a regular population and an empty archive (external set). After the initialization of the population, all the individuals comprising the population are tested for non-dominance. The first non-dominated front is copied and preserved as an elite set in the archive. The size of this elite set may be greater than the archive and hence it may be possible that the same solution may lie in the archive and also the main population. In order to distinguish between the same solutions that are present in both the archive and the main population, two terms are coined: strength of an individual $i, (S_i)$ and fitness of an individual i (F_i).

$$S_i = \frac{n_i \, (number \ \ of \ \ main \ \ population \ \ dominated \ \ by \ \ i)}{1 + N \, (Main \ \ population \ \ size)} \qquad (3.41)$$

$$F_i = 1 + \sum S_{ij} \, (for \ \ all \ \ j \ \ dominating \ \ i) \qquad (3.42)$$

It is followed by the reunion of the main population and the elite in the selection process for recombination and mutation. While strength is assigned to the members present in the archive, fitness is assigned to all the members present in the main population in order to give preference to the member present in the archive during selection for subsequent genetic operations. The method of calculating strength and fitness is given in equations 3.41 and 3.42. In all cases, strength is less than fitness; thus, giving preference to the individual present in the archive and hence helping in preserving the elite. There may be some member in the population that does not dominate each other and very little information is obtained from their domination status. In order to make the search more effective, a clustering algorithm is also used for density estimation. After recombination and mutation, the old population is replaced by the offspring population. This process is repeated until convergence or any other stopping criterion.

This algorithm worked well on quite a few problems; at the same time, there were a few drawbacks too:

1. When there was only one individual in the archive, then all the members of the main population were assigned the same fitness. In such cases, this algorithm behaved like a random search.
2. The clustering algorithm was beneficial for the archive only and not for the whole population. In addition, due to its complexity, its use added to the computational cost.
3. Since the size of the archive was fixed, outer solutions were usually lost, in spite of using the complex clustering algorithm. Hence, it was very difficult to get a good spread of non-dominated solutions.

The methodologies used to address the previous issues have led to the development of a much improved and efficient version of SPEA. This was popularly called the Strength Pareto Evolutionary Algorithm 2 (SPEA2).

Strength Pareto Evolutionary Algorithm 2 is basically an improved version of SPEA and hence it is referred as SPEA 2. The steps involved in the evolution process are as follows:

1. Initialize the population P_0 of size N and set an empty archive set $\overline{P_0}$ of fixed size $\bar{N}$.
2. At any time t, fitness is assigned to all the individuals present in both P_t and $\overline{P_t}$.
3. Thereafter, all the non-dominated individuals that are present in both P_t and

$\overline{P_t}$ are copied to $\overline{P_{t+1}}$. If the size of $\overline{P_{t+1}}$ is greater than the $\bar{N}$, a truncation operator is used to restrict its size to $\bar{N}$ in such a way that it minimizes the loss of individuals on the outer boundary. At the same time, if the size of $\overline{P_{t+1}}$ is less than $\bar{N}$, a few dominated solutions from P_t and $\overline{P_t}$ are copied to the archive in order to maintain its fixed size at $\bar{N}$.

4. In case any stopping criterion is reached, then the set of non-dominated individuals (A) present in $\overline{P_{t+1}}$ at that particular generation are declared the non-dominated set of solutions.

5. In order to fill the mating pool, a binary tournament selection operator is applied at the same time when there is a replacement in $\overline{P_{t+1}}$.

6. After applying the recombination and mutation operator on the mating pool, $\overline{P_{t+1}}$ is set as the new population and the process is repeated again from step (2) until convergence or any other stopping criterion is not encountered.

3.9 INVERSE DESIGN

Here, a user will fix the targeted objective that they want to achieve. Thereafter, they will work on mathematical modeling for determining the parameters that will help them achieve those desired objectives.

In this book, we have dealt with inverse design in a few ways:

- Parallel Coordinate Chart (PCC): Several case studies throughout this book shows how PCC can be used as an effective tool for inverse deign.
- Case study 7: Designed a composition of Ni-based superalloys in the Citrine informatics platform.
- Case study 12: Development of GUI in MATLAB® for estimating various additions in LD-steelmaking by solving a set of equations for material balance.

These are not the conventional approaches reported in the literature on inverse design, but, through these approaches, we are able to determine parameters that proved to be helpful in achieving the targeted properties.

3.10 MULTI-CRITERION DECISION MAKING (MCDM) (MIETTINEN 1999)

A researcher can perform all these simulations and all of the results look promising. In multi-objective optimization, we have dealt with up to ten properties at a time (Case Study 6). It is extremely difficult for any algorithm to optimize all ten objectives and provide with a set of unique solutions, or Pareto set. Many times a user has to sacrifice on one or a few objectives while dealing with such problems.

This is the point, when an expert can be helpful, who has years of experience with that system. An expert can suggest the objective that is extremely important for a certain application, and the objectives where a user can sacrifice to a certain extent, and up to what extent. Mathematically, this concept is known as multi-criterion decision making (MCDM), which was implemented in this work through modeFRONTIER software. In this algorithm, a user assigns weights on a set of objectives and selects a set of potential solutions through the algorithm.

In alloy design, one may get thousands of new candidates through multi-objective optimization. A user has to select a few candidates for manufacture of these alloys (Case Study 6). MCDM can be extremely helpful in such a situation. A rescarcher can set weights on objectives through their understanding of physical metallurgy of these alloys. Then, the algorithm will provide a set of solutions that are expected to meet the requirements set by the research group.

3.11   CASE STUDY #3: DESIGNING NEW MOLECULES/ REFRIGERANTS BY COMPUTATIONAL CHEMISTRY AND AI APPROACH USING SCHRODINGER MATERIALS SCIENCE SUITE SOFTWARE

3.11.1   BACKGROUND

In computational chemistry, SMILES stands for simplified molecular-input line-entry system. The SMILES notation allows the user to portray complex molecules as line notation. SMILES notation may be imported into a number of molecule editors to visualize it in 2D and 3D.

Based on the available literature, a set of 295 classes of refrigerant molecules has been identified in this work (Bhargava 2010).

SMILES notation of one of these 295 refrigerants can be expressed as FC(F)(Cl)C(F)(Cl)Cl, while its representation in 2D and 3D has been shown in Figures 3.5 and 3.6, respectively. We used the Jaguar module in the Schrodinger Materials Science suite software for this work (Bochevarov et al. 2013). These molecules were plotted in the Canvas module of the Schrodinger Materials Science suite software.

These 295 candidates were treated as the training set for a multi-objective design optimization problem, where the variables were the number of bonds on the carbon atom as shown in Table 3.2.

Take an example of a molecule with four carbon atoms like 1-Butanol. It can be represented as CH3-CH2-CH2-CH2-O-H. Thus, the C1 atom will be the atom that is bonded with (OH). So, for C1, there are two C-H bonds, one C-O bond, and one C-C bond. Thus, the X1 will be 1, X9 will be 2, and X25 will be 1. Now, for C2 and C3,

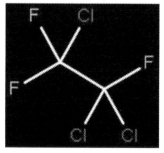

FIGURE 3.5   2-D representation for FC(F)(Cl)C(F) (Cl)Cl, one of the 295 refrigerant molecule in SMILES format.

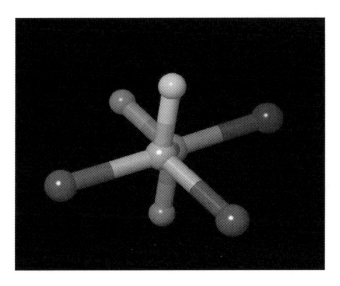

FIGURE 3.6   3-D representation for FC(F)(Cl)C(F)(Cl)Cl, one of the 295 refrigerant molecule in SMILES format.

TABLE 3.2
List of Variables for the Refrigerants: X1 to X28

	C1	C2	C3	C4
Number of C-C bonds	X1	X2	X3	X4
Number of C=C bonds	X5	X6	X7	X8
Number of C-H bonds	X9	X10	X11	X12
Number of C-Cl bonds	X13	X14	X15	X16
Number of C-F bonds	X17	X18	X19	X20
Number of C-Br bonds	X21	X22	X23	X24
Number of C-O bonds	X25	X26	X27	X28

---

there are two C-H bond and two C-C bonds. This means, X2 = 2, X3 = 2, X10 = 2, and X11 = 2. For C4, there are three C-H bonds and one C-C bond; thus, X4 = 1 and X12 = 3. Therefore, for 1-Butanol, Table 3.2 will look as shown in Table 3.3.

This way, a table of 28 variables was created for the 295 refrigerants (Bhargava 2010).

Five objectives are as follows:

P1 = Normal boiling temperatures [Maximize]

P2 = Enthalpy of vaporization [Maximize]

P3 = Tropospheric half-life time [Minimize]

P4 = Root mean squared value of the difference in the vapor pressure of the particular organic compound and that of Freon-12

TABLE 3.3
List of Variables for 1-Butanol

	C1	C2	C3	C4
Number of C-C bonds	1	2	2	1
Number of C=C bonds	0	0	0	0
Number of C-H bonds	2	2	2	3
Number of C-Cl bonds	0	0	0	0
Number of C-F bonds	0	0	0	0
Number of C-Br bonds	0	0	0	0
Number of C-O bonds	1	0	0	0

P5 = Biodegradability obtained from BIOWIN 5

Based on this design optimization problem, the optimizer generated several new molecules. SMILES notation of 43 of these molecules was valid, while hundreds of SMILES were invalid.

3.11.2 GLOBAL WARMING POTENTIAL AND INFRARED ABSORPTION SPECTRA

From an application point of view, the global warming potential is one of the essential parameters for a refrigerant. In order to calculate global warming potential, a researcher needs to perform experiments and determine the infrared spectra of a potential molecule (Etminan et al. 2014).

Performing experiments for such a large data set is a complicated task. In this work, we determined the infrared spectra of these molecules through computational chemistry. We used the Jaguar module of the Schrodinger Materials Science suite for performing ab initio-based calculations. Through the Jaguar module, we were able to determine the infrared spectra of all the molecules available in the previous study.

The intensity of the peaks observed for 295 refrigerants has been plotted in Figure 3.7, while the intensity of 43 optimized refrigerants have been plotted in Figure 3.8.

In Figure 3.8, there is one sharp peak. Apart from that peak, one can observe that the intensity of all the other peaks in Figure 3.8 are lower than that of Figure 3.7. The global warming potential is directly related to the strength with which a given species absorbs infrared spectra.

Since, the intensity of the infrared spectra is lower for the 43 optimized candidates, it means these candidates are expected to have a lower GWP when compared with initial 295 candidates.

SMILE notation of one of the 43 optimized refrigerant molecule can be represented as F/C = C(\Cl)C(F)(F)OCCl. A 2-D representation of this molecule has been shown in Figure 3.9 and a 3-D representation has been shown in Figure 3.10.

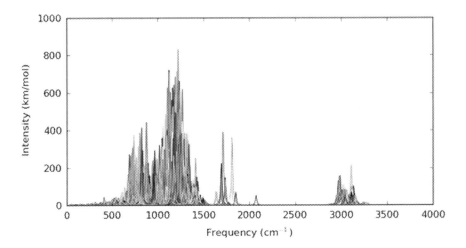

FIGURE 3.7   295 refrigerants: Infrared spectra calculated in Jaguar module of Schrodinger Materials Science Suite software.

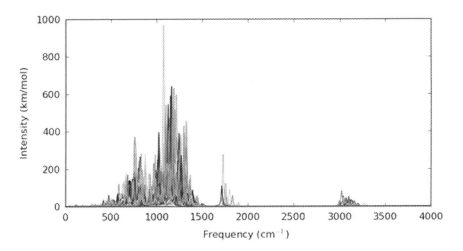

FIGURE 3.8   43 optimized refrigerants: Infrared spectra calculated in Jaguar module of Schrodinger Materials Science Suite software.

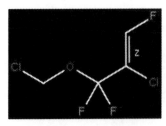

FIGURE 3.9   2-D representation for F/C=C(\Cl)C(F)(F) OCCl, 1 of the 43 optimized refrigerant molecules in SMILES format.

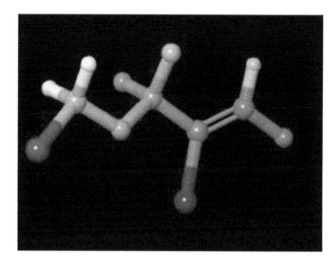

FIGURE 3.10 3-D representation for F/C=C(\Cl)C(F)(F)OCCl, 1 of the 43 optimized refrigerant molecule in SMILES format.

3.11.3 CONCLUSIONS

In this case study, we have demonstrated that computational chemistry software can be helpful in screening new molecules for a specific purpose. We utilized data from a previous work and analyzed it through a commercial software for estimating infrared spectra of all the molecules. One can observe that 43 optimized molecules have less intensity when compared to the initial 295 molecules. Only 1 of the 43 optimized molecules has a sharp peak.

In multi-objective optimization, algorithms can generate hundreds of new SMILES that are invalid. A user must be careful when dealing with complex molecules.

The Schrodinger Materials Science suite is quite popular in the drug design community. Thus, in this book, we included a case study on analyzing patterns from data available from the open literature. In drug design for discovering a drug molecule for a new disease, a researcher analyzes the drugs that have been found effective for a number of similar diseases on various parameters known to medical science community. Thereafter, they formulate their problem statement and start working on new molecules for designing a drug molecule for the new disease.

In this work too, we have analyzed a list of SMILES notations from the open literature. Our problem formulation includes analysis of infrared spectra intensity of all the molecules, both the training set and the 43 optimized molecule obtained through multi-objective design optimization. At this point, we can conclude that the multi-objective design optimization approach was helpful in designing molecules that have a lower intensity for infrared spectra. The absorbance of infrared spectra is directly related with global warming potential (GWP); thus, the 43 optimized molecules are expected to have a lower GWP than the initial set of 295 molecules.

Global warming has been an important research topic for decades. At this point in time, COVID-19 and its random mutations demand new initiatives that can guide researchers in designing new molecules in a short period of time.

Multi-objective design optimization, coupled with computational chemistry, will be helpful for researchers in analyzing potential candidates for a set of targeted properties on a computer. Having vital insights prior to experimentation is extremely helpful, as experimentation and characterization is extremely expensive and time consuming.

4 Case Study #4: Computational Platform for Developing Predictive Models for Predicting Load-Displacement Curve and AFM Image: Combined Experimental-Machine Learning Approach

4.1 INTRODUCTION: BACKGROUND

A nanoindentation test and AFM imaging have been discussed in detail in Chapter 2. In this chapter, we presented an innovative platform for developing AI-based predictive models for indentation depth and AFM image as a function of test parameters like loading rate, maximum applied load and holding time (Jha and Agarwal 2021). The developed models are capable of predicting the indentation depth/load displacement curve and the AFM indentation image for new test conditions/parameters. A user can directly use the data generated by the nanoindentation machine as text (.txt) and HDF (.hdf) and develop predictive models of indentation "depth" and "AFM indent image". We have experimentally verified our approach on two (2) different materials.

The load displacement curve is used for estimating mechanical properties, while AFM indentation imaging is used to study deformation of a material (Oliver and Phar 1992, Liu et al. 2020). This platform will be helpful in a virtual simulation of load-displacement curve along with AFM indent imaging for a large number of new test parameters. Predictive models are developed by processing experimental data obtained from a few experiments. If the tests are performed in a way that the parameters vary a lot, a user can use this platform for even nine experiments. Thus, this platform will be helpful in significantly reducing the number of indents needed for analyzing a material.

DOI: 10.1201/9781003167372-4

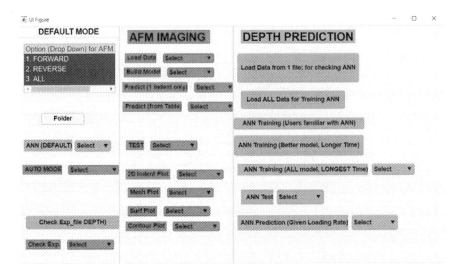

FIGURE 4.1   GUI/APP developed in MATLAB programming language.

The software (GUI/App) has been developed in MATLAB® programming language. It can be used as a MATLAB APP by MATLAB users on any operating system. We have also prepared stand-alone (free) software for Windows OS for users without a MATLAB license (Jha and Agarwal 2021). The GUI/APP is shown in Figure 4.1.

4.2   CASE STUDIES

As mentioned in the previous section, we tested this software on two different coatings. Cited publications will provide detailed information on the nanoindentation experiments (Bakshi et al. 2010, Lahiri et al. 2010, Pitchuka et al. 2014, 2016).

4.2.1   Cold Sprayed Aluminum Based Bulk Metallic Glass (AL-BMG) Coating (Pitchuka et al. 2014, 2016)

The samples were available in "As Sprayed" condition, and then these samples were heat treated. In this chapter, we have presented test results for the "As Sprayed" samples. Analysis of heat-treated samples have been published in another article (Jha and Agarwal 2021). During nanoindentation tests, the maximum applied load was set at 1,000, 2,000, and 4,000 μN. Time to reach maximum load was 5 seconds; that is, 5 seconds for loading and 5 seconds for unloading for all the loads. Holding time was set at 0, 5, 10, 15, 30, 60, 120, and 240 seconds for each of the maximum loads. A total of 24 indentation experiments were performed, out of which AFM indentation imaging was performed in 18 of these indentation tests. Figure 4.2 shows the displacement vs time plot for all the test cases. Polynomial fit equations have been tabulated in Table 4.1. The data in Figure 4.2 can be fitted by a

TABLE 4.1

Polynomial Fit over the Curves Shown in Figure 4.2

Coating	Polynomial Fit	R^2
AS1000	$y = 1E\text{-}05x^3 - 0.0036x^2 + 0.3965x + 7.6794$	0.9759
AS2000	$y = -3E\text{-}06x^3 - 0.0012x^2 + 1.1027x + 12.08$	0.9900
AS4000	$y = -0.0032x^2 + 1.6619x + 1.2826$	0.9822
HT1000	$y = 8E\text{-}06x^3 - 0.0034x^2 + 0.4604x + 4.6456$	0.9798
HT2000	$y = 5E\text{-}06x^3 - 0.0029x^2 + 0.5315x + 7.8413$	0.9991
HT4000	$y = 9E\text{-}06x^3 - 0.0041x^2 + 0.7303x + 11.742$	0.9922

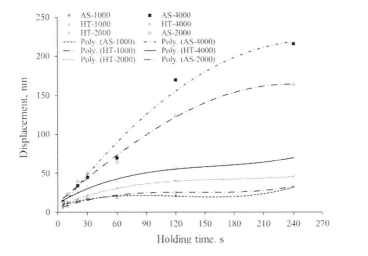

FIGURE 4.2   Displacement vs holding time for AL-BMG coating.

polynomial fit, but each point on Figure 4.2 represents a nanoindentation test. In the latter text, we have described the data set obtained from each of these tests.

Predictive models were developed for the following:

- Indentation depth prediction and plotting load-displacement curve: From a total of 24 indentation experiments, 24 text files were generated.
- Indentation image prediction using imaging files in HDF (.hdf) format: Imaging was performed in both FORWARD and REVERSE directions. Thus, from 18 AFM imaging tests, there were 18 HDF files for each of GF, GR, TF, and TR corresponding to each of these indents.

4.2.2   AL-5CNT COATING (BAKSHI ET AL. 2010)

A total of nine tests were performed. In this case, maximum applied load varied from 984–5,936 μN. Time to reach maximum load was fixed at 10 seconds; that is,

10 seconds for loading and 10 seconds for unloading. Holding time was fixed at 5 seconds for all the tests, but loading rate varied from about 98–594 μN/s.

- Indentation depth prediction and plotting load-displacement curve: For developing models, nine machine generated text files were used.
- Indentation image prediction using imaging files in HDF (.hdf) format: Imaging was performed in only the FORWARD direction. There were nine sets of HDF files for each of GF and TF corresponding to each of these indents. In this case, the number of image files was not sufficient for developing predictive models. It is advised to use more imaging files for more accurate and reliable prediction.

4.3 RESULTS: APPLICATION OF MACHINE LEARNING ALGORITHMS

4.3.1 CASE STUDY 4.2.1: LOAD-DISPLACEMENT CURVE PREDICTION FOR AL-BMG AS SPRAYED/CAST COATING

For developing predictive models for indentation depth, machine-generated text (.txt) files have been used. Each text (.txt) file contains about 1,034 row, along with some text, blank rows, and some rows where the sensor picked up negative load and displacement data. These text files contain data under the column headers for Depth, Load, and Time, respectively, in adjacent columns for a combination of test parameters. Any negative data, blank rows and text need to be removed as it will affect model development. This platform preserves usable data and automatically removes unusable information while importing of data from the text (.txt) file. In this case study, we identified "maximum applied load" and "holding time" as the test parameters that affect the indentation depth. Additionally, data under the "Load" and "Time" column in the text files was also identified as parameters that affect the data recorded under indentation "Depth" column in the text files generated by the machine.

ANN model for indentation "Depth" was developed as a function of "Maximum applied Load", "Holding time", "Load", and "Time".

4.3.1.1 ANN Model Development in This Work

From the experiments, we obtained 24 text files from 24 indents. Out of these, 23 files were used during model development, and one was kept for comparison. From 23 files, a large data set of 23,420 rows was extracted after initial preprocessing. This data is can be used for developing AI-based predictive models using the concept of artificial neural network (ANN), which usually needs a large data set for training. Deep Learning Toolbox™ (formerly Neural Network Toolbox™) in MATLAB 2019 (MATLAB 2019b) was used for developing an ANN model for indentation depth. Data was divided as follows: 70% of the data was used for training (16,394 samples), 15% for validation (3,513 samples), and 15% for testing (3,513 samples). Neurons in the hidden layer were varied between 10 and 20. In this work, we have presented results using an ANN model with one hidden layer consisting of 15 neurons. In MATLAB, one has the option of choosing an algorithm for training, and we chose Levenberg-Marquardt as the training algorithm. The ANN model was trained multiple times, until the platform provided us with a model that has an error similar to or lower than the training and validation test. Error values have been tabulated in Table 4.2. ANN models are prone to overfitting; thus, we selected a model that provides us with a load-displacement curve for a new test condition that is acceptable by someone who performs experiments. Catching the trend of the load-displacement curve is more important. Error metrics have been tabulated in Table 4.2, where MSE stands for mean square error, which is the average of the squared differences between the outputs and targets. A lower MSE value corresponds to a lower error. R means regression or relationship between model output and targets. An R value near 1 means a close relation, while an R value of 0 means a random relation. Figure 4.3 shows the comparison between the load-displacement curve obtained through an ANN model and experiments for As-sprayed sample of cold sprayed aluminum based bulk metallic glass (AL-BMG (AS)) coating.

TABLE 4.2
Error Metrics for ANN for Depth AL-BMG (as Sprayed)

	Samples	MSE	R
Training	16,392	4.54912	0.99952
Validation	3,511	4.58923	0.99943
Testing	3,511	4.58476	0.99978

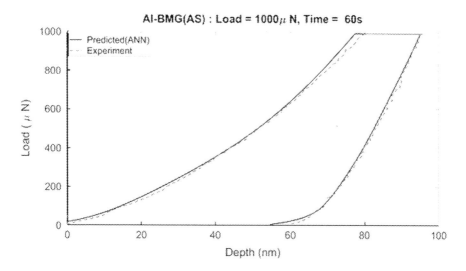

FIGURE 4.3 Comparison: Experiment vs. predicted for load of 1,000 μN and holding time of 60 seconds for AS-Sprayed aluminum based bulk metallic glass (AL-BMG (AS)) coating.

4.3.2   Case Study 4.2.1: AFM Image Prediction for AL-BMG as Sprayed/Cast Coating

As indicated, four HDF files are generated: GF, GR, TF, and TR. In summary, AFM imaging is performed in two modes, "forward" and "reverse". We had 18 sets of files (4 HDF) corresponding to each of these tests. Seventeen sets of files (4 HDF files each) were used during model development, and one set of files (4 HDF) was kept for comparison.

> *Predictive model for "AFM indent image" was developed as a function of "Maximum applied Load", and "Holding time" using data stored in the AFM imaging files in HDF (.hdf) format.*

Figure 4.4 shows the comparison between the images obtained from the machine (4.4(a) and (b)), and that predicted by our approach (4.4(c) and (d)) for a case for which data was not introduced to the model.

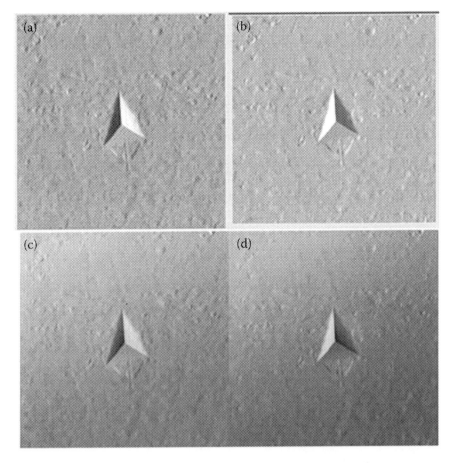

FIGURE 4.4   (a) and (b): System-generated image; (c), (d): Predicted image (current work) for maximum applicable load of 2,000 μN and 30 seconds of holding time.

From Figure 4.4 one can observe that the predicted indent looks similar to the indent generated from the machine. From the indentation depth model, we already have the expected load-displacement curve. Thus, the image of our predicted indent can prove to be useful as it captures the morphology of the indent observed from the experiments.

Figure 4.5 shows the 3-D topographical image of the indent predicted through our software for maximum applied load of 2,000 μN and 30 seconds of holding time.

Figure 4.6 shows the predicted images for new test conditions: maximum applied load was 1,800 μN and holding time was 65 seconds. As mentioned earlier, it takes a while to predict the indentation image.

In Figure 4.6, one can observe that the predicted AFM image looks realistic. This new test parameter was specified by the user. This platform provides flexibility to the user in analyzing the AFM image in the form of 2-D indent image, and 3-D indent profile instantaneously. A user is required to provide some experimental AFM image files in HDF format and set new test conditions. Within an hour, a user can easily simulate an AFM image for several new test conditions.

4.3.3 Case Study 4.2.2: Load-Displacement Curve Prediction for Al-5CNT Coating

A total of nine text files were obtained, of which eight were used to develop models. One of the files was used for comparative purposes.

ANN model for indentation "Depth" was developed as a function of "Maximum applied Load", "Loading rate", "Load", and "Time".

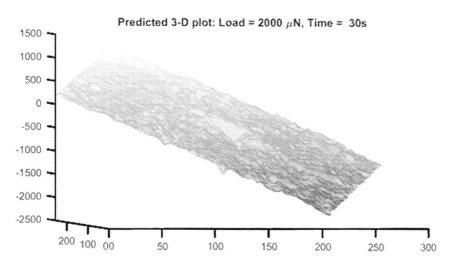

FIGURE 4.5 Predicted topographical image of indent for load 2,000 μN; holding time 30 seconds.

(a)

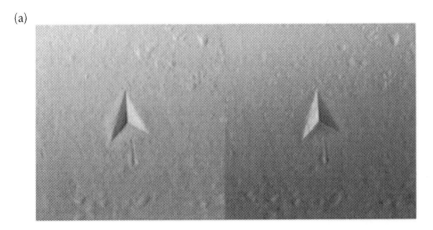

(b)

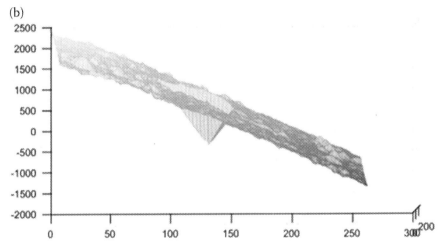

FIGURE 4.6 Predicted 2-D and 3-D topographic image for (a) Sl. No. 3: Load: 1,800 μN, Time: 65 seconds.

Figure 4.7 shows the load-displacement curve comparison between experiments and ANN model for Al-5CNT, while error metrics for an ANN model have been tabulated in Table 4.3.

Thereafter, we chose a new set of test conditions: maximum load = 3,500 μN and loading rate = 350 μN/s. We generated a loading curve or load and time reading in an Excel sheet. In the "loading curve" or "load" and "time" reading generated by the machine, we have approximately 1,034 rows or 1,034 sets of "load" and "time" readings and corresponding "depth" in the text (.txt) file generated by the machine. In Excel, we generated data which varies linearly, and the distribution was different when compared with the readings from the machine.

We used this new set of generated data in Excel and tested our ANN model. Figure 4.8 shows the load-displacement curve predicted by our model for a new set of test conditions, maximum load of 3,500 μN and loading rate of 350 μN/s. We can

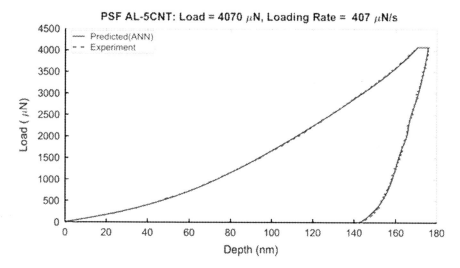

FIGURE 4.7 Comparison: Experiment vs. predicted for maximum load of 4,070 μN, loading rate of 407 μN/s, and holding time of 5 seconds for Al-5CNT.

TABLE 4.3
Error Metrics for ANN for Depth AL-5CNT

	Samples	MSE	R
Training	6,370	0.4096	0.99993
Validation	1,365	0.5466	0.99991
Testing	1,365	0.4249	0.99993

observe that the load-displacement curve predicted by our ANN model is able to capture trends shown by a load-displacement curve generated by experiments.

Our intention was to check if our ANN model for "depth" is able to predict "depth" or provide a load-displacement curve for a new set of readings of load and time variation. This is to check if the curve looks similar to the machine-generated load-displacement curve for a new set of readings, so that the ANN model can be used as a predictive tool for screening prior to performing experiments. From Figure 4.8, we can visually see that the model performs well and the generated load-displacement curve looks similar to a traditional load-displacement curve generated from actual experiments even when we used a completely different set of load and time variation.

4.3.4 CASE STUDY 4.2.2: AFM IMAGE PREDICTION FOR AL-5CNT COATING

Figure 4.9 shows the AFM image generated from experimental HDF files and plotted by our software for a maximum load of 4,689 μN, loading rate of 469 μN/s, and holding time of 5 seconds for an Al-5CNT sample.

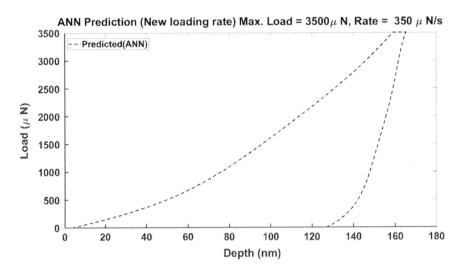

FIGURE 4.8   Load-displacement curve predicted through ANN model for maximum load of 3,500 μN, loading rate of 350 μN/s, and holding time of 5 seconds for Al-5CNT.

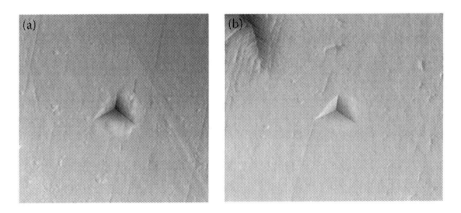

FIGURE 4.9   Experimental holding time of 5 s for Al-5CNT: maximum load: (a) 4,689 μN, (b) 4,070 μN, and loading rate: (a) 469 μN/s and (b) 407 μN/s.

4.3.5   HA CNT

Hydroxy apatite coatings are applied on a range of biomedical implants (Lahiri et al. 2010). In this case, an HA-CNT coating on a titanium implant was analyzed through nanoindentation. We possessed only one set of calculations, and the AFM image was missing. So, we used our software for analysis of these raw files.

In this example, we have plotted the load-displacement curve (Figure 4.10) and AFM indent image (Figure 4.11) through our software.

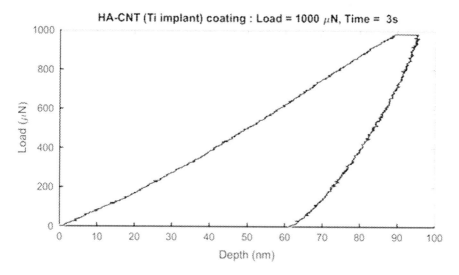

FIGURE 4.10 Load-displacement curve for HA-CNT coating on titanium implant.

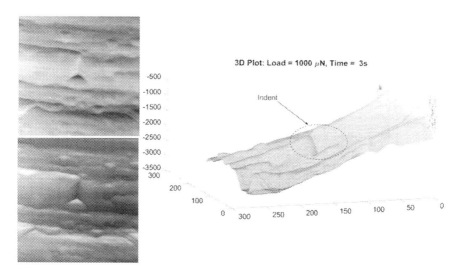

FIGURE 4.11 AFM image plotted for HA-CNT coating on titanium implant.

4.4 CONCLUSIONS

In this case study, we demonstrated the efficacy of using our software (GUI/APP) on data generated by a nanoindentation test. Raw data are in text and HDF format. This software is capable of predicting a load-displacement curve, which is essential for determining mechanical properties of a material. Additionally, one can visualize the deformation of materials for new test conditions by predicting an

AFM image through our platform. The predicted image looks realistic when compared with an actual AFM image. Predicting an image takes a few minutes, but load-displacement curves can be estimated instantaneously. This software does not need information about material properties. This software can be utilized as a screening tool for testing new experimental test conditions prior to performing indentation tests.

5 Case Study #5: Design of Hard Magnetic AlNiCo Alloys: Combined Machine Learning-Experimental Approach

5.1 INTRODUCTION: DESCRIPTION OF CASE STUDY AND GOALS

A brief discussion on various magnetic terminologies has been discussed in Chapter 2, along with a section on hard magnetic AlNiCo-based alloys.

In this chapter, we will focus on the following:

- Data generation and acquisition
- Multi-objective problem formulation
- Results: Supervised machine learning
- Results: Unsupervised machine learning
- Discussion: Advantages and disadvantages of applying AI algorithms over physical models and experiments

5.2 DESIGN VARIABLES (COMPOSITION), TARGETED PROPERTIES, METAMODELING, AND MULTI-OBJECTIVE OPTIMIZATION PROBLEM FORMULATION

5.2.1 DATA ACQUISITION

This work was performed in cycles to check upon improvements. Concentration of alloying elements was changed over the course of this study. Variable bounds, or upper and lower limit of the composition of alloying elements, have been tabulated in Table 5.1. Cycle number, alloys included in those cycles, and the best alloy per cycle have been tabulated in Table 5.2. Regarding performance, all the alloys were experimentally tested for properties listed in Table 5.3. In Table 5.3, there are ten objectives listed. An alloy that performs better on a number of properties in a given cycle has been referred to as the "best alloy" in Table 5.2.

Composition of alloys in cycle 1 (1–80) and cycle 4 (91–110) were generated through Sobol's algorithm. Thus, in cycle 1, 80 alloys were manufactured and tested. Based on the results, five alloys were predicted by artificial intelligence

DOI: 10.1201/9781003167372-5

Artificial Intelligence-Aided Materials Design

TABLE 5.1
Upper and Lower Limits for Alloying Elements Included in this Study of AlNiCo-type Alloys (Jha 2016)

	Variable Bounds (weight percent)		
Alloying elements	1–85	86–143	144–180
Cobalt (Co)	24–40	24–38	22.8–39.9
Nickel (Ni)	13–15	13–15	12.35–15.75
Aluminum (Al)	7–9	7–12	6.65–12.6
Titanium (Ti)	0.1– 8	4–11	3.8–11.55
Hafnium (Hf)	0.1–8	0.1–3	0.095–3.15
Copper (Cu)	0 – 6	0–3	0–4.5
Niobium (Nb)	0–2	0–1	0–1.5
Iron (Fe)	Balance to 100%		

TABLE 5.2
Design Cycles and Alloy Numbers (Jha 2016)

Cycle Number	Number of Alloys Designed	Best Alloy (number)
1	1–80	30
2	81–85	84
3	86–90	86
4	91–110	95
5	111–120	117
6	120–138	124
7	139–143	139
8	144–150	150
9	151–160	157
10	161–165	162
11	166–173	169
12	174–180	180

(AI)–based approaches and experimentally tested, which is part of cycle 2. Based on alloys 1–85, five more alloys were predicted through AI-based approaches and experimentally tested, which is part of cycle 3. At the end of cycle 3, the research group decided to change the bounds on alloying elements. Thus, in cycle 4, again Sobol's algorithm was used to generate the composition of 20 new alloys. The research group checked the progress in each cycle and then decided the composition bounds and means of generating new compositions.

Compositions were either generated through Sobol's algorithm, a quasi-random number generator, or the compositions were generated through an AI-based approach. The AI-based approach included developing predictive models through a

TABLE 5.3

Quantities to Be Simultaneously Optimized Using Multi-Objective Optimization

Properties	Units	Objective
Magnetic energy density ($(BH)_{max}$)	$Kgm^{-1}s^{-2}$	Maximize
Magnetic coercivity (H_c)	Oersted	Maximize
Magnetic remanence (B_r)	Tesla	Maximize
Saturation magnetization (M_s)	Emu/g	Maximize
Remanence magnetization (M_r)	Emu/g	Maximize
$((BH)_{max})$/mass	$m^{-1}s^{-2}$	Maximize
Magnetic permeability (m)	$KgmA^{-2}s^{-2}$	Maximize
Cost of raw material	\$/Kg	Minimize
Intrinsic coercive field ($_jH_c$)	Am^{-1}	Maximize
Density(ρ)	Kgm^{-3}	Minimize

supervised machine learning approach for the properties listed in Table 5.3. It was followed by multi-objective optimization of various properties through evolutionary or genetic algorithms (Table 5.3). This resulted in a large pool of new compositions, in the order of hundreds to thousands in some cases. It is extremely difficult to choose a set of alloys from this group for experiments. The unsupervised machine learning approach was then used to determine the best candidates from this group. In modeFRONTIER software, we applied the concept of multi-criterion decision making (MCDM) to select a few alloys. A set of new compositions was predicted in each cycle for experimental validation.

5.2.2 TARGETED PROPERTIES AND MULTI-OBJECTIVE PROBLEM FORMULATION (JHA 2016)

Magnetic properties that are essential for application were identified and have been listed in Table 5.3. In Table 5.3, there is a column objective and this is the basis of multi-objective problem formulation. A total of eight properties are maximized, while the alloys are expected to have a low cost and lower density; thus, two properties are to be minimized.

It is a complicated and challenging task, as any optimization approach can efficiently handle two to three objectives while performing optimization. Thus, in this work, our group utilized a number of statistical and artificial intelligence–based approaches to handle this complex case of alloy design and optimization.

5.2.3 SOFTWARE USED IN THIS WORK

- ESTECO modeFRONTIER: Used for developing models using both supervised and unsupervised machine learning algorithm. Multi-objective optimization was performed using a number of approaches listed in Chapter 3.

- Sigma Technologies (IOSO software): Used for developing surrogate models and multi-objective optimization of targeted properties.
- IBM SPSS: Unsupervised machine learning algorithm principal component analysis (PCA).
- WEKA: Data analysis.
- KNIME: Data analysis.

5.3 APPLICATION OF SUPERVISED MACHINE LEARNING ALGORITHMS

As stated in Chapter 3, it is advised to analyze data through various statistical tools and then proceed towards applying AI-based algorithms. Supervised machine learning algorithms can be used for developing predictive models, but each algorithm has its advantage and limitation. Magnetic properties and surrogate models that best fit these properties are listed in Table 5.4. All of these models were developed in ESTECO modeFrontier software. In this book, we have used an artificial neural network (ANN) and deep learning approach, but in Table 5.4, these algorithms are missing. The reason is that the ANN approach requires a large data set, while we have a table of a maximum of 180 alloys. Initially, we started with only 80 alloys, which is not sufficient enough for developing surrogate models through an ANN approach.

For an identical processing route, it is important to determine the composition-property relationship. A variable (alloying element) can directly influence an objective (property); that is, increasing the amount of a particular variable leads to an increase in value of that property. A relation can be "inverse"; that is, the increase in value of a variable results in a reduction in value of a property. A relation can be termed "nil", which means there is no effect of that variable on that property. Since we are dealing with a noisy and complex data set, the relation can be a mix of all of

TABLE 5.4
Surrogate Models Selected for Targeted Objectives (Jha 2016)

Properties	Surrogate Models
$(BH)_{max}$	RBF (Gaussian)
H_c	RBF(MR)
B_r	RBF(IMQ)
M_s	ED
M_r	RBF(IMQ)
$(BH)_{max}$/mass	AKR(Gaussian)
magnetic permeability	RBF(IMQ)
cost of raw materials	RBF(MQ)
jH_c	AKR(Gaussian)
density	RBF(MQ)

TABLE 5.5

Single Variable Response for Magnetic Properties (Objectives) (Jha 2016)

Objectives	Variable Response							
	Fe	Co	Ni	Al	Ti	Hf	Cu	Nb
$(BH)_{max}$	Nil	Nil	Mix	Nil	Nil	Nil	Nil	Nil
H_c	Mix	Mix	Mix	Inv	Mix	Dir	Dir	Mix
B_r	Mix	Mix	Mix	Inv	Mix	Dir	Dir	Inv
M_s	Dir	Inv	Dir	Mix	Inv	Dir	Mix	Mix
M_r	Nil	Nil	Nil	Nil	Nil	Nil	Nil	Nil
$(BH)_{max}$/mass	Nil	Nil	Nil	Nil	Nil	Nil	Nil	Nil
Magnet permeability	Mix	Mix	Mix	Mix	Inv	Mix	Mix	Mix
cost of raw material	Inv	Inv	Inv	Dir	Dir	Dir	Inv	Dir
jH_c	Mix	Mix	Mix	Inv	Inv	Mix	Dir	Mix
density	Mix	Dir	Mix	Inv	Inv	Mix	Mix	Dir

them, as can be seen in most of the cases in Table 5.5. In mix mode, a variable may have not have affected an objective for a particular composition, which may directly or inversely affect that objective for other compositions.

5.4 MULTI-OBJECTIVE OPTIMIZATION FOR DETERMINING NOVEL COMPOSITION OF NEW MAGNETS

At this stage, we had ten supervised machine learning models for ten properties mentioned in Table 5.3. Table 5.3 also provides with a list of properties. A total of eight (8) of these properties are maximized and two of these properties are minimized. Multi-objective optimization was performed in modeFRONTIER software through several concepts of evolutionary algorithms or genetic algorithms.

This approach was followed for each of the 12 cycles. We have mentioned previously that new compositions were either generated through Sobol's algorithm or the strategy mentioned in this section.

Figure 5.1 shows the parallel coordinate chart for alloys included in cycles 1–8. Three properties have been shown on the plot, $(BH)_{max}$, B_r, and H_c. All the lines linked with cycle 1 are at the bottom of the PCC plot. One can observe that alloys included in cycles 2–8 perform better than alloys included in cycle 1. Cycle 1 was the initial batch of 80 alloys. Alloys included in cycles 2–8 were developed by learning over the initial batch of 80 alloys in subsequent generations.

Figure 5.2 shows the plot between H_c and B_r, where H_c is in the negative direction. On a traditional B-H curve, (BH) max is calculated on the negative side of the B-H curve. Thus, we plotted all 180 alloys on this figure. In Figure 5.2, one can observe that the best alloy so far was alloy #124, developed through modeFRONTIER software. A few of the best-performing alloys have been marked by a number. In Figure 5.2, the alloys were differentiated by the approach from which composition of these candidates was generated. The word "Experiment" on

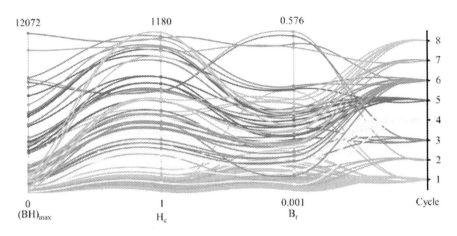

FIGURE 5.1 Parallel coordinate plots showing $(BH)_{max}$, B_r, and H_c values for cycles 1–8.

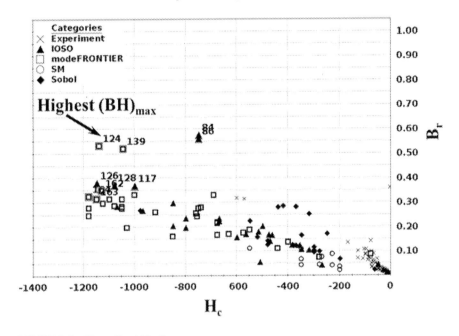

FIGURE 5.2 H_c vs B_r: 180 alloys.

Figure 5.2 corresponds to the initial set of 80 alloys in cycle 1. All of these alloys were experimentally tested. From Figure 5.2, one can observe that the initial set of alloys was almost non-magnetic. Even then, by applying a proper design strategy and monitoring progress in small batches, we were able to significantly improve upon the properties shown in Figure 5.2.

In Figure 5.3, we have plotted all the 180 alloys on a copper vs hafnium plot. In Figures 5.1 and 5.2, we have explained that the alloys generated in cycle 2 and above were superior in performance in comparison with the alloys included in cycle 1.

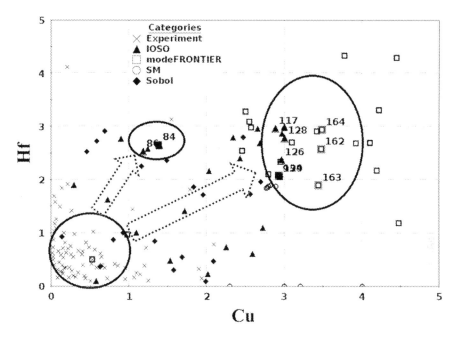

FIGURE 5.3   Cu vs Hf: 180 alloys.

In each cycle, we predicted new compositions and tested them, moved ahead with the same strategy, or used a different approach. But this work primarily involved determining new compositions that will be helpful in improving multiple properties of the alloy.

Regarding composition, there are eight alloying elements. In Figure 5.3, one can observe the composition regime of alloys generated through various methods. The composition regime of alloys in cycle 1 and the best-performing alloys have been marked in Figure 5.3. One can observe that the composition regime is completely different. It will be impossible to alter the composition of an eight-element alloy in this manner.

This demonstrates that one can significantly improve upon multiple properties by just changing the composition while the processing method remains the same. Initial 80 alloys were almost non-magnetic, while alloy #124 was the best-performing alloy in this study. Our approach includes multi-objective design-optimization and monitoring progress in small batches. This strategy helped in significantly improving upon multiple properties in a few cycles only.

5.5   APPLICATION OF UNSUPERVISED MACHINE LEARNING ALGORITHMS

In this work, three algorithms were used: principal component analysis (PCA), hierarchical clustering analysis, and self-organizing maps (SOM) (Jha et al. 2016,

SOM Component plot: 180 Alloys

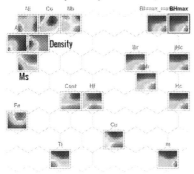

FIGURE 5.4   SOM component plots for 180 alloys.

2017). In this chapter, we will only discuss the SOM analysis (Jha et al. 2017). Figure 5.4 shows the SOM component plot for all 180 alloys included in this work.

In Figure 5.4, one can observe that $(BH)_{max}$ and $(BH)max\_mass$ are clustered together. $(BH)max\_mass$ in this figure means $(BH)max/mass$. Br and M_r are clustered together. H_c and $_jH_c$ are the inverse of each other in terms of magnetic property and in Figure 5.4, these two units are close to each other. Thus, the SOM algorithm is able to detect several correlations that have been reported on magnetic alloys.

Thereafter, each of these cells (in Figure 5.4) were individually analyzed. A few of the cells have been discussed in this chapter. For each of these cells, we estimated the best matching unit and the results have been tabulated in Table 5.6. In Table 5.6, one can observe that units 48, 56, and 57 contain all the best alloys observed in this study. All the figures will be explained with respect to Tables 5.4 and 5.6.

Figure 5.5 shows the distribution of $(BH)_{max}$ on the SOM plot. Units 48, 56, and 57 are clustered together. These units are associated with the highest values of $(BH)_{max}$ observed in this work.

Figure 5.6 shows the distribution of B_r on the SOM plot. Units 48, 56, and 57 are associated with the highest values of B_r observed in this work.

Figure 5.7 shows the distribution of H_c on the SOM plot. Units 48, 56, and 57 are associated with the highest values of H_c observed in this work.

Similar plots were plotted for all ten properties. We observed that seven of the objectives have been maximized. Three objectives that were not optimized are

TABLE 5.6
SOM Analysis: Best Matching Units

Unit #	Candidate Alloys
48	95, 125, 147, 165
56	111, 114, 117, 126, 127, 128, 143, 161, 162, 163, 164, 166, 169, 180
57	84, 86, 124, 139

SOM Component plot: (BH)max

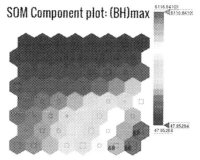

FIGURE 5.5 SOM plot showing distribution of $(BH)_{max}$.

SOM Component plot: Br

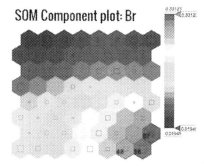

FIGURE 5.6 SOM plot showing distribution of B_r.

SOM Component plot: Hc

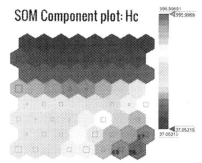

FIGURE 5.7 SOM plot showing distribution of H_c.

density, saturation magnetization (M_s), and cost. For alloys in units 48, 56, and 57, density and saturation magnetization (M_s) are average, while cost is high.

Figure 5.8 shows the distribution of Fe content on the SOM plot. Fe content for alloys included in units 48, 56, and 57 are in the vicinity of the lowest values of Fe content reported in this work.

Bounds on properties for alloys included in unit 57, along with SOM prediction, have been reported in Table 5.7. Variable bounds on alloy composition for alloys included in unit 57, along with SOM prediction, have been reported in Table 5.8. Tables 5.7 and 5.8 are to be used for guidance as these are average compositions. For each unit cell, one can observe a square of varying size is

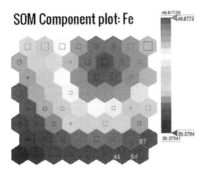

SOM Component plot: Fe

FIGURE 5.8 SOM analysis for Iron content (wt. %).

TABLE 5.7
SOM Component Analysis for Properties of Unit #57

Properties	Maximum	Minimum	SOM Prediction
$(BH)_{max}$	12072	8358	6116
H_c	1140	750	957.82
B_r	0.576	0.519	0.331
M_s	130	111	126.57
Mr	66	57	37.02
$(BH)max\_mass$	1.90E8	9.44E7	5.92E7
μ	3.32E-4	6.67E-6	1.46E-4
Cost	3051	3033	2974
$_jH_c$	90718	59683	76220
Density	7255	6835	7142

TABLE 5.8
Component Analysis of Elements for Unit #57

Element	Maximum	Minimum	SOM Prediction
Al	8.7424	7.2055	7.236
Co	36.888	35.7158	36.366
Cu	2.939	1.1804	2.62
Fe	32.4409	31.921	32.659
Hf	2.6559	2.068	2.119
Nb	0.931	0.6295	0.801
Ni	13.824	13.544	13.395
Ti	4.762	4.125	4.807

inscribed. The larger the size of the square, the larger is the number of candidates that are positioned on its vertices. SOM algorithm is best known for topology preservation; thus, similar alloys are positioned together, and this similarity is based on SOM parameters (not on physical metallurgy). Thus, these alloys are similar, and all of these alloys are not the best alloys. Additionally, we are dealing with a complex system of eight elements, a simple average over the composition values for a set of alloys is not the best approach for determining a novel composition. SOM analysis is used for understanding patterns. SOM maps were used for sorting a few candidates from a list of hundreds to thousands of candidates obtained from multi-objective optimization.

5.6 COMPARISON OF EXPERIMENTAL AND PREDICTIONS FROM AI-BASED APPROACHES WITH PHYSICAL MODELS

5.6.1 COMPARISON WITH COMMERCIAL ALLOYS

Alloy #124 is the best alloy developed in this work. This alloy was compared with a few commercial alloys (Table 5.9). In Table 5.9, one can observe for alloy 124 that B_r is low; otherwise, H_c is comparable.

Alloy 124 was developed through multi-objective design optimization. For all the properties mentioned in Table 5.3, predictive models were developed through supervised machine learning algorithms. Now, supervised machine learning algorithm's performance depends on the training set, the data over which the models are trained. Alloy 124 belonged to cycle 6. Thus, the training data set for the predictive models contained only a few alloys that belong to top ten overall alloys, while the majority of the alloys are below par.

This shows that it is possible to improve upon properties of a material through multi-objective design optimization even if the training set contains mostly

TABLE 5.9

Chemical Composition of the Best Optimized Alloy and Several Commercial AlNiCo Alloys (Jha 2016, Palasyuk et al. 2013)

Fe	Co	Ni	Al	Ti	Hf	Cu	Nb	$(BH)_{max}$	H_c	B_r
Composition (Wt %)								$J\ m^{-3}$	Oe	Tesla
Chemical composition of alloy #124										
32.33	36.86	13.54	7.2	4.1	2.06	2.94	0.93	12072	1140	0.532
Chemical composition of the commercial alloy AlNiCo 5-7										
49.9	24.3	14.0	8.2	0.0	0.0	2.3	1.0		740	1.35
Chemical composition of the commercial alloy AlNiCo 8										
30.0	40.1	13.0	7.1	6.5	0.0	3.0	0.0		1860	0.82
Chemical composition of the commercial alloy AlNiCo 9										
35.5	35.4	13.1	7.0	5.0	0.0	3.2	0.5		1500	1.06

substandard solutions. Even a few nice solutions will be helpful in generating candidates with superior properties in subsequent generations, iterations, or cycles.

5.6.2 Comparison: AI Predictions vs Experiments

So far, we have demonstrated the efficacy of multi-objective design optimization, which is great, as until alloy #80; most of the alloys were nonmagnetic. Thus, optimizing 7 out of 10 properties, and generating a novel composition in a few trials is exceptional.

But, in similar research, it is rarely mentioned how far the algorithm's initial prediction is from its experimentally validated value. This analysis is a bit different but is important. Mathematical and statistical tools had an advantage over traditional experiments that were based on experience or mere intuition of the researcher. Similarly, AI-based algorithms have some advantages over traditional mathematical and statistical models.

We have demonstrated the scope of further improvement. Theoretical calculations also suggest that a $(BH)_{max}$ of 20 MGOe is achievable, while in literature it is reported that AlNiCo has a $(BH)_{max}$ of about 10 MGOe.

So, every research demonstrates improvement and scope of improvement in the future. In recent years, uncertainty quantification is an important topic. Thus, a detailed analysis of uncertainty associated with these models is important. In Chapter 1, we have performed calculations to address uncertainty quantification within the framework of the CALPHAD approach. In this chapter, we have demonstrated prediction uncertainty through plots and explained it through text.

Figure 5.9 shows the comparison between AI-based predictions for $(BH)_{max}$ and their experimental validation. This plot is interesting as usually predictive models

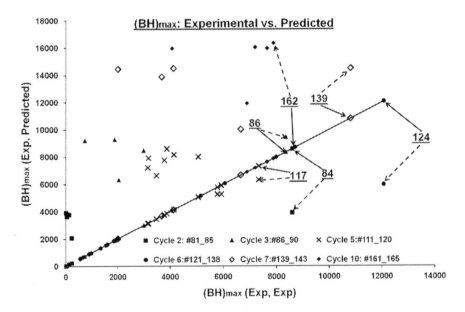

FIGURE 5.9 $(BH)_{max}$: AI prediction vs experimental validation.

overestimate the objective, and experimental validations have a lower value. In Figure 5.9, one can observe a completely different scenario. For alloy 86 and 117, model predictions are close to experimental validation. The models overestimated for 86, and underestimated for 17, but the difference is marginal. For alloy 84, the model significantly underestimated the $(BH)_{max}$ values. Alloys 84, 86, and 117 are IOSO predictions. Alloys 124 and 139 were predicted by modeFRONTIER. Software underestimated the value of 124, while overestimated the value of 139.

Figure 5.10 shows the comparison between AI-based predictions for H_c and its experimental validation. The model predictions for H_c are far more accurate when compared with $(BH)_{max}$. For H_c, the difference between AI-based predictions and experimental validation is small. AlNiCo-based magnets are magnetized in the final stages of heat treatment. H_c is an intrinsic property and it depends on materials anisotropy. If one takes into consideration the complexity of this problem, they will accept the model predictions and the associated error.

Figure 5.11 shows the comparison between AI-based predictions for B_r and its experimental validation.

In Figure 5.11, models developed in different software are associated with some level of uncertainty. The performance of predictive models in quantifying Br is similar to the performance of predictive models in quantifying $(BH)_{max}$.

Thus, H_c models are acceptable, whereas models for B_r and $(BH)_{max}$ need a further parameter tuning. Models for B_r and $(BH)_{max}$ are uncertain as they sometimes perform as per our expectation, while they overestimate and underestimate the values for most of the candidates.

Every model is associated with an error or uncertainty. For AlNiCo-based magnets, theoretical calculations show that a value of 20 MGOe is achievable for

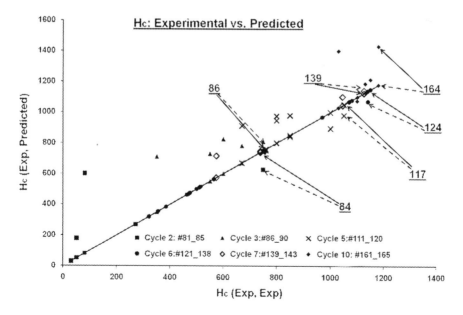

FIGURE 5.10 H_c: AI prediction vs experimental validation.

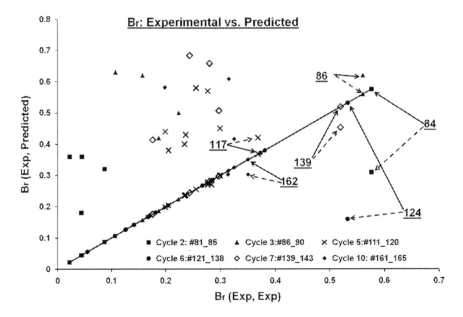

FIGURE 5.11 B_r: AI prediction vs experimental validation.

$(BH)_{max}$, but the best-known AlNiCo-based magnets have a $(BH)_{max}$ at about 10 MGOe. This means that theoretically calculated, $(BH)_{max}$ is twice the amount of experimentally verified $(BH)_{max}$. One has to trust theoretical calculations based on concepts of physics if they are working on magnets.

Thus, most important is to strike a balance and look at possibilities of improving predictive capabilities of various models. Be it a theoretical model or AI-based model, both can be efficiently used while performing experiments.

5.7 CONCLUSIONS

It totally depends on the reader to believe only the literature on magnets that explains it in terms of basic physics. They can totally go along with traditional alloy design; that is, working around the composition of a known alloy and change the composition of an element at a time. Or they go along with an AI-based approach. All of these approaches have advantages and limitations. All of them are associated with significant error. The AI-based method has the advantage over them, with respect to time frame; that is, the time frame and the number of trials involved in the AI-based approach is significantly less when compared to discovering a new alloy based on concepts of physics or time frame in comparison with determining a novel composition with multiple improved properties through traditional alloy design. Time frame of experimentation, including manufacturing and testing of all of the 180 alloys was about 2 years.

In this work, we have demonstrated the efficacy of multi-objective design optimization, which is great, as until alloy #80, most of the alloys were non-magnetic. After developing 90 alloys, we decided to generate random compositions again. So, composition generated for the initial 110 alloys contained 100 candidates with random composition. Most of the alloys' performances were below par and not per our expectation. Our best alloy is alloy #124. So, through proper monitoring and multi-objective design optimization, within 15 trials, we were able to generate the composition of the best alloy reported in this work. In alloy design, this is absolutely remarkable as 15 trials is an extremely small number as there are 8 elements and 10 properties involved. Improving upon properties and generating a new composition takes decades through traditional experiments. Through traditional experiments, it will be extremely difficult to design a completely new composition and optimize multiple properties simultaneously. Thus, optimizing 7 out of 10 properties, and generating a novel composition in 15 trials, is exceptional.

6 Case Study #6: Design and Discovery of Soft Magnetic Alloys: Combined Experiment-Machine Learning-CALPHAD Approach

6.1 INTRODUCTION

Magnetic terminologies have been described in detail in Chapter 2. We are dealing with a FINEMET-type soft magnetic alloy with alloying elements Fe, Si, Nb, Cu, and B.

In this chapter, we will describe the modeling and simulation and experimental results.

- Atomistic simulations: Simulations were performed for determining magnetic states of various phases in the Fe-Si system. Simulations were carried out using ab-initio/DFT-based software VASP. VASP stands for Vienna Atomistic Simulation Package, while DFT stands for Density Functional Theory. For visualization, a software "Vesta" was used. We have described how this software can be helpful in determining atomic position and planes, especially for magnetic (Ferro and Antiferro) calculation.
- CALPHAD simulation: In this work, the Fe_3Si phase (α'' D03) phase was identified as the desired phase through information obtained from atomistic simulations and from literature. The Fe-Si phase diagram was plotted to determine the region on the phase diagram where the Fe_3Si phase (α'' D03) is stable. The Fe_3Si phase (α'' D03) is a metastable phase, and it is not directly observed on the phase diagram. Thereafter, the G-X curve, or Gibbs energy vs composition curve, was plotted so that a reader can visualize the calculated G-X curve, which is usually presented in schematic form in literature. Finally, precipitation kinetics simulations were carried out under the framework of the CALPHAD approach for simulating nucleation and growth of the Fe_3Si phase (α'' D03) phase or nanocrystals for a class of FINEMET-type alloys.
- Artificial intelligence: We used both supervised and unsupervised machine learning (ML) approaches in this work.

DOI: 10.1201/9781003167372-6

- Optimizing composition-processing-structure: The supervised ML approach was used for developing predictive models for mean radius and volume fraction of the Fe_3Si phase (α'' D03) phase as a function of composition, annealing time, and isothermal annealing temperature.
- New magnet discovery: The unsupervised ML approach was used for new magnet discovery. The concept of self-organizing maps (SOM) was used for determining pattern in data collected from the literature. Thereafter, supervised ML approaches were used for developing models for saturation magnetization (B_s) and coercivity (H_c) as a function of composition. Finally, new compositions were predicted for new magnets by multi-objective optimization through a set of evolutionary algorithms (EAs) or genetic algorithm (GAs).
- Experiments: We have presented information on calibrating CALPHAD-based model parameters. CALPHAD model predictions are in good correlation when compared with experimental results analyzed through atom probe tomography (APT).

6.2 AB-INITIO DFT: ESTIMATION OF MAGNETIC STATES FOR VARIOUS PHASES

In this section, we present an estimation of magnetic states for various phases associated with the Fe-Si system. Calculations have been tabulated in Table 6.1.

In Table 6.1, one can observe that both Ferro and Antiferro states were considered. In Fe-Si system, iron (Fe) is the magnetic phase and silicon (Si) is nonmagnetic. Calculation of magnetization for the Ferro state is simple as magnetic moments needs to be aligned in one direction only. After the simulation, a user can obtain the total magnetic moment of the crystal and magnetic moment per atom for the Fe atom.

Calculation of magnetization for the Antiferro state is not that simple as magnetic moments needs to be aligned in the opposite direction for adjacent atoms in the unit cell. Additionally, the number of moments in the positive Z-direction must be equal to the number of moments in the negative Z-direction. Take for example Fe_3Si; this crystal is not suitable for Antiferro simulation, as there are odd Fe atoms, so magnetic moments in the positive Z-direction will not balance magnetic moments in the negative Z-direction. In Table 6.1 and Figure 6.1, it can be observed that the number of atoms per unit cell is 16. This way, we will have a system of 16 elements, where 12 atomic positions will be occupied by Fe atoms and 4 atomic positions will be occupied by Si atoms (Figure 6.1).

For Fe_3Si, there are 16 atoms and magnetic moments need to be placed in the opposite direction on 12 Fe atoms. This is a challenging task if a user wants to determine atomic positions manually. A user will need to write a computer code, or it can be done through an open-source visualization software like VESTA (Momma and Izumi 2011). VESTA is the easiest way is to find the atomic planes and put a value of (+1) and (−1) for consecutive atomic planes. This will be particularly helpful for larger unit cells like $Fe_{11}Si_5$, where we considered 128 atoms per unit

TABLE 6.1
Estimation of Magnetic States for Various Phases in the Fe-Si System

Crystal /Phase	Structure (No. of Atoms/ Unit Cell) {Space Group}	Energy without Entropy (ev)				FERRO (Magnetization) (Bohr Magneton, μ_B)		ANTIFERRO (Magnetization) (Bohr Magneton, μ_B)	
		FERRO		ANTIFERRO		TOTAL (μ_B)	Per Fe Atom (μ_B)	TOTAL (μ_B)	Per Fe Atom (μ_B)
		TOTAL	Per Atom	TOTAL	Per Atom				
Fe_3Si (Stable)	DO3 (16) {Fm-3m}	−128.253	−8.016	−125.876	−7.867	19.915	1.660	−0.000	−0.000
$Fe_{11}Si_5$ (Unstable)	B2 (128) {Pm-3m}	−1005.743	−7.857	−1002.338	−7.831	104.188	1.184	−1.888	−0.021
$FeSi_2$ (Stable)	A2 (48) {Cmce}	−330.280	−6.881	−330.280	−6.881	0.000	0.000	−0.000	−0.000
$Fe2Si$ (Stable)	(6) {P2$_1$3}	−46.937	−7.822			4.083	1.021		
$FeSi$ (Stable)	B19 (8) {P2$_1$3}	−59.52	−7.44	−59.632	−7.454	4.056	1.014	0.000	0.000

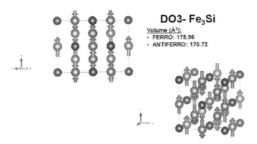

FIGURE 6.1   Unit cell for Fe$_3$Si (D03), where magnetic moments on adjacent Fe atoms are aligned in opposite direction.

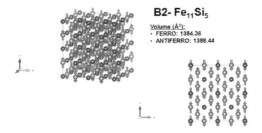

FIGURE 6.2   Unit cell for Fe$_{11}$Si$_5$ (128 atoms), where magnetic moments on adjacent Fe atoms are aligned in the opposite direction.

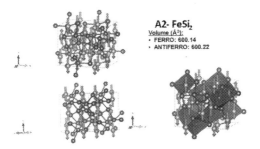

FIGURE 6.3   Unit cell for FeSi$_2$ (48 atoms), where magnetic moments on adjacent Fe atoms are aligned in the opposite direction.

cell, of which 88 are Fe and 40 are Si in Figure 6.2. Similarly, we have plotted FeSi$_2$ in Figure 6.3, Fe$_2$Si in Figure 6.4, and FeSi in Figure 6.5.

6.3   CALPHAD: THERMODYNAMICS/PHASE TRANSFORMATION AND KINETICS

Based on atomistic simulation and information from the literature, the desired phase for the current system is Fe$_3$Si (α'', D03). Thus, we studied the given class of the FINEMET-type system under the framework of the CALPHAD approach. We chose a composition (wt. %) and plotted the phase diagram as shown in Figure 6.6.

FIGURE 6.4 Unit cell for Fe_2Si (6 atoms), where magnetic moments on adjacent Fe atoms are aligned in the opposite direction.

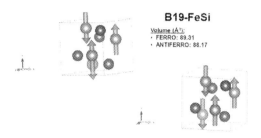

FIGURE 6.5 Unit cell for FeSi (8 atoms), where magnetic moments on adjacent Fe atoms are aligned in the opposite direction.

6.3.1 IDENTIFICATION OF STABLE AND METASTABLE PHASES (JHA ET AL. 2017)

Following phases were observed in Figure 6.6: BCC_A2 or α-Fe, B2_BCC (Ordered and Disordered, M3SI (or Fe_3Si)(FE:SI), M2B_TETR (or Fe_2B)(Fe: B rich phase), BM (or FeB) (B:FE), FE3B (or Fe_3B) (FE:B phase), MB_B33 (or FeB) (FE:B), M23C6 ($Fe_{23}B_6$) (FE:B phase), FE2SI (or Fe_2Si)(FE:SI phase), M5SI3 (or Fe_5Si_3) (FE:SI), FCC_A1 (FE:CU), CUZN_EPSILON (CU:VA) (observed in) DIAMOND_FCC (Si) and LAVES_PHASE_C14 (Fe: Si, Nb rich phase). Notations like BCC_A2, B2_BCC are the notations included in the Thermo-Calc database, while the companion notations are notations that we observed in the literature on a FINEMET-type alloy and the Fe-Si system in particular.

Out of these phases, M3SI closely resembles the Fe_3Si phase; thus, we analyzed the Fe-Si system for confirmation in this regard.

The equilibrium phase diagram for the Fe-Si system was plotted (Figure 6.7), but one cannot observe the M3SI phase on this diagram.

In the next calculation, we removed the B2_BCC phase and then performed the equilibrium calculation and plotted the phase diagram for the Fe-Si system (Figure 6.8). Still, the M3SI phase cannot be observed in Figure 6.8.

Thus, in the next calculation, the BCC_A2 phase was suppressed or removed while performing an equilibrium calculation and the Fe-Si phase diagram was plotted, as can be seen in Figure 6.9. In Figure 6.9, one can observe that the M3SI phase can be observed on the phase diagram. From literature on the Fe-Si system, we confirmed that

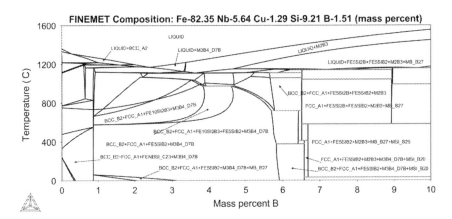

FIGURE 6.6 Phase diagram for a given composition (wt. %) of FINEMET-type soft magnetic alloy.

the M3SI phase region in Figure 6.9 coincides with the phase region of the Fe_3Si (D03) phase when compared with literature available on the Fe-Si system.

In the CALPHAD framework, stability of a particular phase is studied through Gibb's energy vs composition diagram, or G-X curve. A schematic view of the G-X curve is reported in literature to explain the stability of a phase with respect to composition fluctuation. In this chapter, we performed calculations to plot the G-X curve through Thermo-Calc software. To start with, we combined Figures 6.7, 6.8, and 6.9; the combined view is presented in Figure 6.10.

Gibb's energy vs composition diagram, or G-X curve, is plotted for a constant temperature. Thus, we chose a temperature of 540°C in Figure 6.10 for analyzing various phases that can be stabilized at 540°C.

From these calculations, Gibb's energy vs composition diagram, or G-X curve, was plotted, as can be seen in Figure 6.11.

6.3.2 PRECIPITATION KINETICS SIMULATION (JHA ET AL. 2017, 2018)

After identifying the Fe_3Si phase on the phase diagram, we worked on simulating precipitation kinetics for nucleation and growth of the M3SI phase in the TC-PRISMA module of Thermo-Calc. The composition of the alloy is as follows: $Fe_{72.89}Si_{16.21}B_{6.90}Nb_3Cu_1$ in atomic %. For desired soft magnetic properties, the alloys must possess Fe_3Si grains in a size range between 10 and 15 nm diameter or 5–7.5 nm mean radius. Volume fraction must be above 70% for minimizing magnetostriction.

In the TC-PRISMA module, governing equations is defined by the Kampman-Wagner-Numerical (KWN) approach, which has been described in detail in Chapter 2. For precipitation kinetics simulation, one of the most important parameters is inter-facial energy between the precipitate and the matrix. In this case study, precipitate is M3SI (Fe_3Si), while the matrix is the "amorphous phase". Isothermal annealing simulations were performed at 490, 500, 510, 520, 530, 540, and 550°C. Interfacial

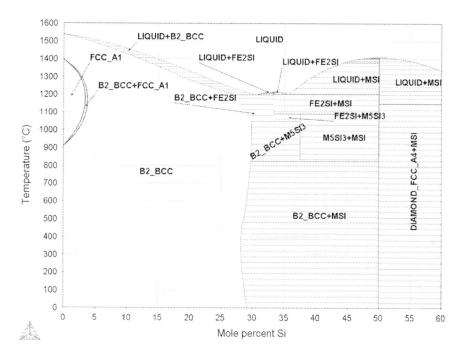

FIGURE 6.7   Fe-Si phase diagram showing the B2_BCC phase.

energies at these temperatures have been tabulated in Table 6.2. Annealing was performed for 2 hours.

Variation of mean radius of M3SI (Fe₃Si) nanocrystals over 2 hours' annealing time for various temperatures has been plotted in Figure 6.12. In Figure 6.13, we have plotted the size distribution of M3SI (Fe_3Si) nanocrystals at 1 and 2 hours. In Figure 6.14, the variation of volume fraction has been plotted. The variation of matrix composition is shown in Figure 6.15.

6.4   APPLICATION OF MACHINE LEARNING ALGORITHMS IN OPTIMIZING PRECIPITATION KINETICS OF DESIRED Fe_3Si PHASE (A″ WITH D03 STRUCTURE) (JHA ET AL. 2018)

From Figures 6.12, 6.13, and 6.14, one can observe that CALPHAD predictions are as per expectation; that, is we were able to achieve mean radius between 5 and 7.5 nm and a volume fraction of about 70% for all the isothermal annealing temperature (490–550°C).

FINEMET composition was provided by NASA-GRC and it can be written as follows:

- Weight %: 82.35Fe- 5.64Nb-1.29Cu- 9.21Si- 1.51B.
- Atomic %: 72.89Fe- 3.0Nb-1.0Cu- 16.21Si- 6.90B.

FIGURE 6.8 Fe-Si phase diagram showing the BCC_A2 phase.

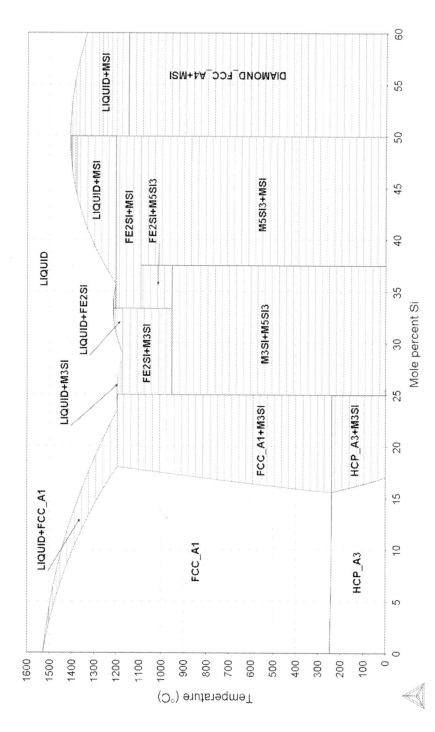

FIGURE 6.9 Fe-Si phase diagram showing the M3SI phase.

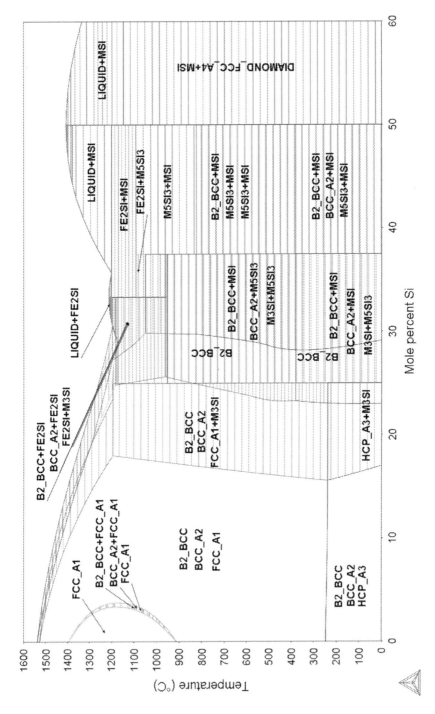

FIGURE 6.10 Combined Fe-Si phase diagram showing B2_BCC, BCC_A2, and M3SI phases.

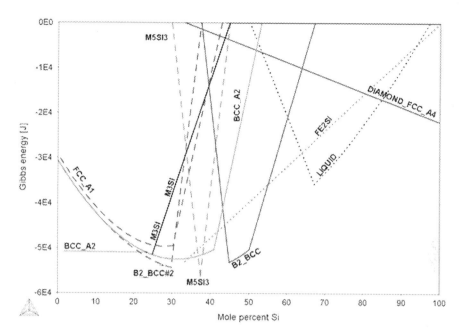

FIGURE 6.11 Gibb's energy vs composition diagram, or G-X curve, for the Fe-Si system at 540°C.

TABLE 6.2

Interfacial Energy between the M3SI (Fe_3Si) Phase and Amorphous Matrix as a Function of Temperature for an Alloy of Composition $Fe_{72.89}Si_{16.21}B_{6.90}Nb_3Cu_1$ in Atomic %

Temp. (°C)	σ (× 10^{-2} J/m^2) (Interfacial Energy)
490	8.9716
500	8.9997
510	9.0285
520	9.0582
530	9.0887
540	9.1198
550	9.1516

For this composition, we generated data for mean radius and volume fraction by varying Fe and Si in the vicinity of the mentioned composition. This was done by introducing a parameter X, where X varies from -3 to 3 atom % (−3, −2, −1, 0, 1, 2, 3) so the new alloy composition will be as follows: $Fe_{72.89+x}Si_{16.21-x}B_{6.90}Nb_3Cu_1$ (where

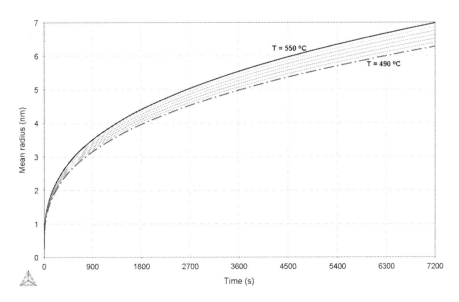

FIGURE 6.12 Mean radius (nm) of M3SI (Fe$_3$Si) vs time (hours) for isothermal annealing between 490 and 550°C.

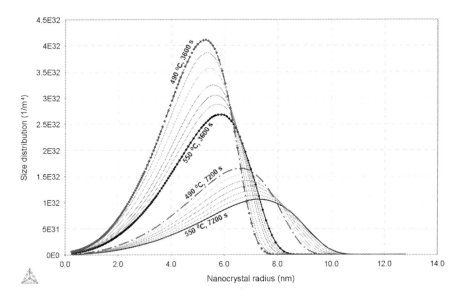

FIGURE 6.13 Size distribution of M3SI (Fe$_3$Si) at 1 and 2 hours for isothermal annealing between 490 and 550°C.

$x = \pm\, 3$ atomic %). Figure 6.16(a) shows the variation of mean radius of M3SI (Fe$_3$Si) for the new compositions over 2 hours of isothermal annealing at 490°C, while Figure 6.16(b) shows the variation of volume fraction. Corresponding interfacial energies has been listed in Table 6.3.

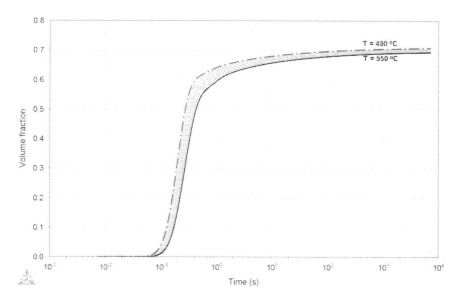

FIGURE 6.14 Volume fraction of M3SI (Fe₃Si) vs time (hours) for isothermal annealing between 490 and 550°C.

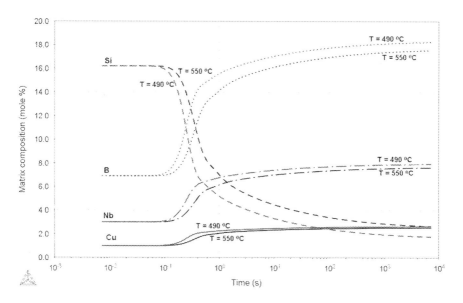

FIGURE 6.15 Variation of matrix composition over course of annealing.

In this work, we used a supervised ML algorithm, K-Nearest algorithm (Jha et al. 2018), a well-known classifier to develop metamodels for mean radius and volume fraction as a function of X (composition), annealing temperature, and annealing time.

(a)

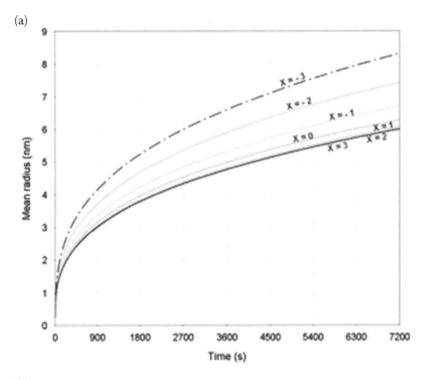

(b)

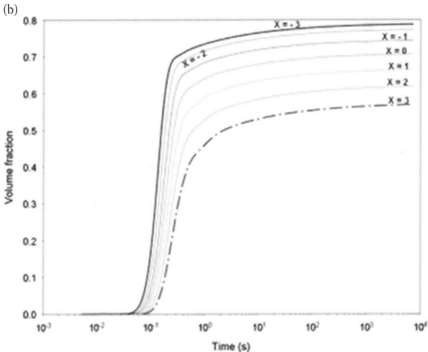

FIGURE 6.16   Variation of mean radius and volume fraction of M3SI (Fe_3Si) crystals for new compositions for isothermal annealing at 490°C.

TABLE 6.3

Interfacial Energy (σ) Values for Isothermal Annealing at 490°C for FINEMET Alloy of Composition Fe72.89+xSi16.21–xB6.90Nb3Cu1 (where x = Ł} 3 Atomic %)

x (Atomic %)	σ (× 10^{-2} J/m^2) (Interfacial Energy) at 490°C
–3	8.1583
–2	8.7147
–1	8.9811
0	8.9716
1	8.7955
2	8.5487
3	8.2837

KNN models for mean radius and volume fraction were quite accurate. Model predictions were in good agreement of CALPHAD-based predictions for new compositions, annealing time, and temperature (Jha et al. 2018). As a result, 20,000 new parameters were generated for compositions, annealing time and temperature, through Sobol's algorithm (Jha et al. 2018). These models could predict mean radius and volume fraction of all these new parameters in less than 2 minutes. In this chapter, we have shown bubble plots for mean radius (Figure 6.17) and volume fraction (Figure 6.18). In these plots, about 300 new sets of parameters were chosen, and the mean radius was predicted through KNN models in Figure 6.17, and volume fraction was predicted in Figure 6.18. In Figures 6.17 and 6.18, the color of the bubbles represents mean radius and volume fraction, respectively, while the diameter of the bubbles is a representation of annealing temperature. The X axis is for composition variation denoted by "X", while the Y axis is for annealing time.

By analyzing Figures 6.17 and 6.18, one can observe the temperature, composition, annealing temperature, and time that needs to be followed for achieving M3SI (Fe$_3$Si) nanocrystals in desired size range and volume fraction.

6.5 EXPERIMENTAL VALIDATION

CALPHAD-based model predictions depend on a range of parameters, and estimation of interfacial energy is extremely complex within the framework of the CALPHAD approach. In this section, we have presented ways of optimizing parameters in CALPHAD-based models, so that CALPHAD-based predictions can be experimentally verified.

Two case studies will be presented in this section to demonstrate the benefits of parametrizing CALPHAD-based models. By changing a few parameters, a CALPHAD-based model can capture trends shown by advanced diagnostic tools like atom probe tomography (APT).

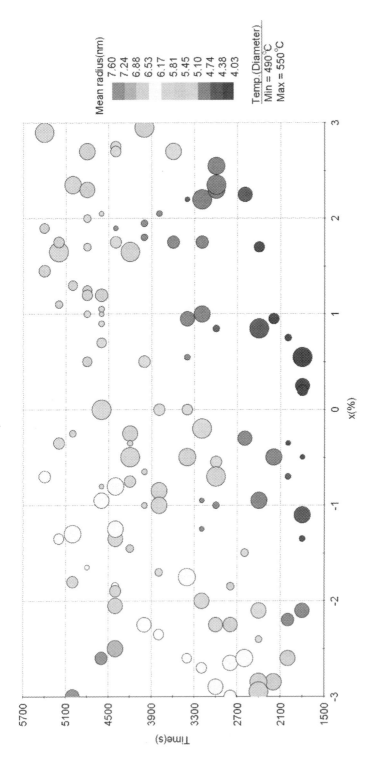

FIGURE 6.17   Bubble plot for mean radius predicted through KNN models for new parameters.

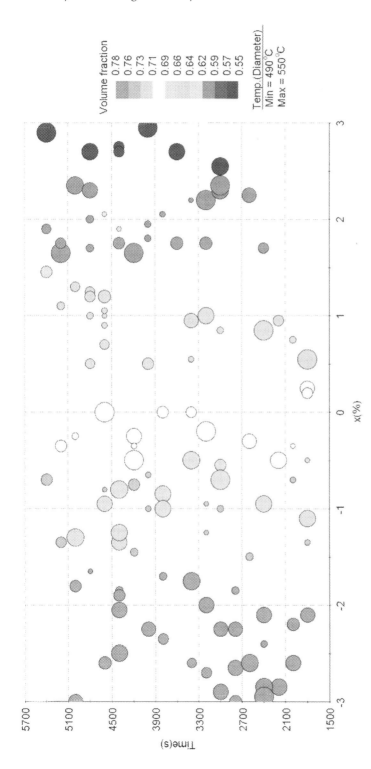

FIGURE 6.18 Bubble plot for volume fraction predicted through KNN models for new parameters.

6.5.1 Isothermal Annealing at 540°C for 1 Hour (Jha et al. 2019)

Size distribution of FCC copper clusters are as follows:

- Experiments (APT): 0.25 nm–3.75 nm
- CALPHAD: 0.26 nm–3.7 nm

6.5.1.1 CALPHAD Model Parametrization

Default value of 0.64 J/m^2 calculated by the software will result in an abnormally large cluster size.

Interfacial energy was varied between 0.27 and 0.80 J/m^2. Optimum interfacial energy was estimated at 0.54 J/m^2. The mean diameter of clusters is around 5.4 nm (mean radius = 2.7 nm). The number density of copper clusters is of the order of 10^{23}/m^3 (Total: 1.5801×10^{23}/m^3). Both mean diameter and number density values predicted by CALPHAD-based models have been reported in literature for FINEMET-type alloys. A comparison between size distribution of clusters observed through APT and predicted through CALPHAD-based models has been shown in Figure 6.19. Detailed analysis has been reported in another article (Jha et al. 2019).

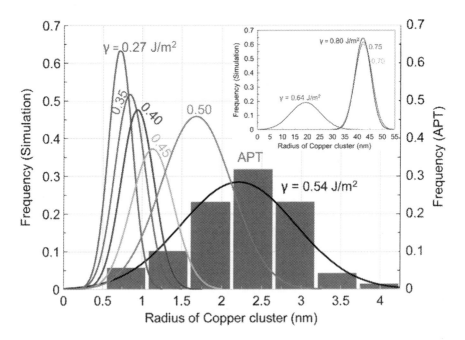

FIGURE 6.19 Comparison: Atom probe tomography (APT) vs CALPHAD-based simulations for various values of interfacial energy (Jha et al. 2019).

6.5.2 ANNEALING TIME OF 0.1 SECONDS (DIERCKS ET AL. 2020)

In this work, we made an attempt to simulate the laser-material interaction (analyzed through APT) through a precipitation model developed by our group under the framework of the CALPHAD approach. In our previous work, we were able to successfully simulate the nucleation and growth of copper clusters during isothermal annealing at 540°C through this model by optimizing the interface energy between the amorphous matrix and copper clusters.

What makes this work novel is the time scale. In the previous work, annealing time was 1 hour, whereas in this current work, cumulative annealing time is 0.1 seconds.

Assumptions made during simulation in TC-PRISMA module of Thermocalc are as follows:

1. In APT, median cluster size and cluster density were estimated at different depths from the surface or the spot of laser material interaction. Temperature at the laser material interaction spot is the maximum (~800 K) and temperature decreases with depth to about 700 K at 200 nm depth (Figure 6.20). Thus, we chose five annealing temperatures for performing isothermal annealing for 0.1 seconds though a precipitation model: 673 K (400°C), 723 K (450°C), 753 K (480°C), 773 K (500°C), and 800 K (527°C). Here, we assumed that the highest temperature reached at the spot of laser material interaction (800 K), while the temperature at 200 nm depth as 673 K.

2. In our previous works, we calculated the molar volume of the amorphous phase $(5.913 \times 10^{-6}$ m^3/mole) from the density of FINEMET ribbon (8.35 g/cm^3). Additionally, we optimized a parameter named mobility enhancement prefactor

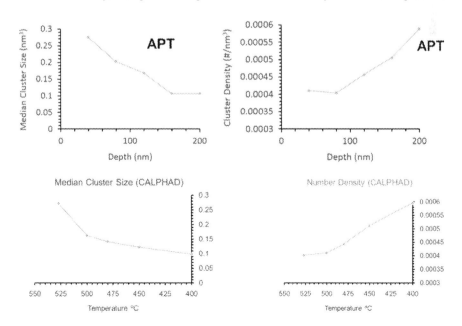

FIGURE 6.20 Comparison between APT characterization and CALPHAD prediction.

and interface energy between copper cluster and amorphous matrix. Mobility enhancement prefactor, a parameter used in Thermocalc, was parametrized to check diffusivity in the amorphous phase, as we assumed the LIQUID phase in the TCFE9 database as an amorphous phase.

3. In this work, too, we optimized the mobility enhancement prefactor and interface energy between the copper cluster and amorphous matrix for each of the mentioned temperatures. CALPHAD predictions were compared with APT results so that CALPHAD predictions are in the vicinity of the median cluster size and cluster density estimated through APT.

4. We observed that with the increase in temperature, both mobility enhancement prefactor and interface energy decreases. The mobility enhancement parameter varied between 2.90E-07 (800 K) and 1.08E-06 (673 K), while the interface energy varied between 0.503 J/m^2 (800 K) and 0.532 J/m^2 (673 K). Model parameters have been tabulated in Table 6.4.

Thus, we were able to successfully simulate the precipitation kinetics of evolution of copper clusters within the framework of the CALPHAD approach. CALPHAD predictions are in the vicinity of results reported through APT, as shown in Figure 6.20.

CALPHAD model parameters listed in Table 6.4 were plotted versus temperature in degrees Celsius and in Kelvin. With temperature in Kelvin, both the parameters show a nice trend which can be easily fitted by a polynomial fit of degree 3. Variation of the mobility enhancement prefactor along with temperature in Kelvin is shown in Figure 6.21, while a similar plot for interface energy (σ in J/m^2) is shown in Figure 6.22. In both Figures 6.21 and 6.22, we have added the polynomial of degree 3 and the R^2 value.

Throughout this book we have mentioned the importance of statistical tools, and in Chapter 3 we have mentioned a simple technique like "curve fitting" can be used for developing predictive models. AI-based algorithms are helpful, but it is nice to start with a simple technique, statistical analysis prior to devoting time in choosing an algorithm for developing surrogate models using an AI-based algorithm.

TABLE 6.4

Model Parameters from CALPHAD Approach

Temp. (°C)	Mobility Enhancement Prefactor	Interface Energy (σ in J/m^2)	Number Density of Copper Cluster (nm^{-3})	Median Cluster Size
400	1.08E-06	0.532	0.000596	0.098568
450	5.45E-07	0.53	0.000512	0.12249
480	3.95E-07	0.529	0.000444	0.141437
500	3.40E-07	0.519	0.00041	0.162755
527	2.90E-07	0.503	0.000402	0.27308

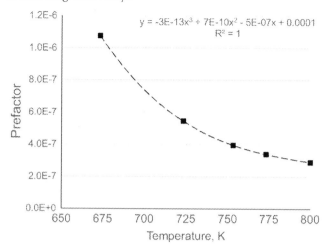

FIGURE 6.21 Mobility enhancement prefactor vs temperature in (K), along with a polynomial fitting trendline.

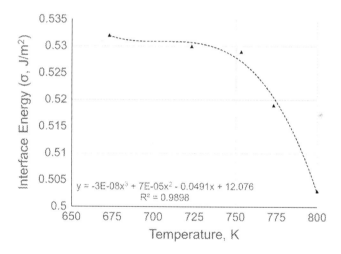

FIGURE 6.22 Interface energy (σ in J/m^2) vs temperature in (K), along with a polynomial fitting trendline.

A reader can analyze Figures 6.21 and 6.22, and use them for calibrating CALPHAD model parameters when they want to analyze a complex system where the matrix phase is the amorphous phase. Mobility enhancement prefactor is a measure of diffusivity, and for the amorphous phase diffusivity will be low. In regular CALPHAD-based analysis, prefactor is set at around 1–5, but these values will not work for assuming LIQUID phase as an amorphous phase. The only way is to decrease diffusivity, which can be achieved by lowering the mobility enhancement prefactor value.

APT is an advanced diagnostic tool, which is both expensive and time consuming. Once parametrized, the CALPHAD approach is quick and the average simulation time for each calculation in this work took about 12–30 seconds. In this work, we observed that the CALPHAD predictions are in the vicinity of the APT results. Additionally, both the mobility enhancement prefactor and interface energy decrease with increasing temperature or we can observe a trend with temperature. Thus, the CALPHAD approach can be used as an effective tool in conjunction with APT analysis. All a user needs is proper selection of model parameters. Once these parameters are optimized, a user can use these CALPHAD-based models for simulating conventional heat treatment where annealing time is in hours. Additionally, a user can perform non-conventional heat-treatment simulations where cumulative annealing time is 0.1 seconds, which is usually not performed through a CALPHAD approach. Through this case study, we demonstrated that a user can not only simulate such heat treatment, but also verify the results through atom probe tomography (APT).

6.6   DISCOVERY OF NEW SOFT MAGNETIC ALLOYS THROUGH THE AI APPROACH

6.6.1   Unsupervised Machine Learning

To start with, we collected a large amount of data from literature on soft magnetic alloys. We used the concept of self-organizing maps in this work. The SOM algorithm is discussed in brief in Chapter 2, so we will present the results here. Topological error for these models was minimized and we chose a model with error 0.012 in the present analysis.

Our purpose of using SOM maps can be listed as follows:

- Find correlations between various variables and properties that can be supported from the literature.
- Classify the data set in various clusters and identify the units/clusters with candidate alloys with a set of superior properties.
- This way a user can use SOM maps for predicting the chemical composition of targeted alloys for superior properties.
- This approach can be used as an additional screening tool for selecting a set of alloys to be manufactured in the next alloy design cycle.

Figure 6.23 shows the component chart, that is distribution of variables (elements) and properties (H_c and B_s) on the SOM hexagonal grid. Here, similar component maps are placed in adjacent positions, making it easy to spot correlations between the properties and variables.

Few important observations are as follows:

- Fe, B_s, H_c, and B are clustered together.
- B_s and H_c are correlated.
- Fe affects B_s.

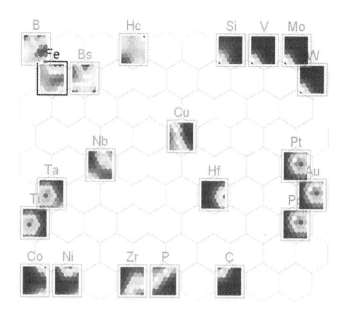

FIGURE 6.23   SOM component plot: Map variables and properties in 2-D space.

- Fe-B system is a known glass former.
- Pt, Au, and Pd are clustered together and have been used in soft magnetic alloys in place of copper for facilitating nucleation.
- Ti and Ta have been used in soft magnetic alloys in place of Nb.
- Co and Ni are clustered together. Both Co and Ni have been used by researchers to substitute iron.

Even though a data set has lots of missing points, we were able to observe a few correlations.

The SOM map is divided into 48 hexagonal units. Eighty-three candidate alloys were used for SOM analysis. The size of the square inscribed in these units is an

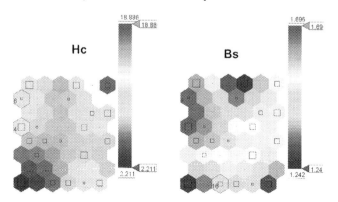

FIGURE 6.24   SOM plot for H_c and B_s.

indication of the number of candidates included in that hexagonal unit. Adjacent units on the SOM maps are better correlated. Thus, any new alloys that lies on these hexagons are expected to have similar properties.

In Figure 6.24, we have highlighted the units for analyzing candidate alloys with a minimum H_c (Units 0 and 16) and maximum B_s (Units 4 and 6). This will be helpful in understanding the later figures where we have plotted similar figures for distribution of critical elements for discovering a new class of soft magnets.

Distribution of various elements in SOM space: Figure 6.25 shows the distribution of iron (Fe) in the SOM space. We have plotted the labels for the

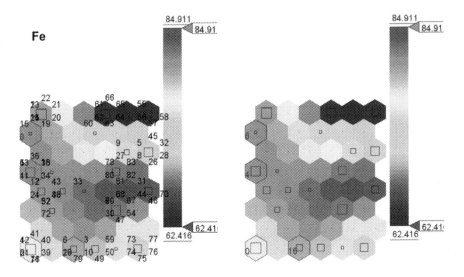

FIGURE 6.25 Distribution of iron in the 2-D space.

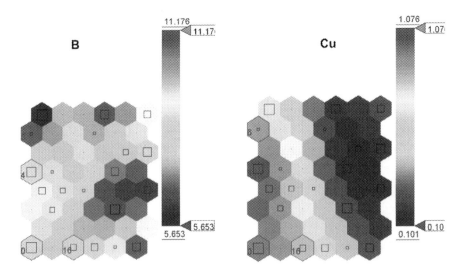

FIGURE 6.26 Distribution of boron and copper in the 2-D space.

candidates on the left plot, while for clarity, we have only highlighted the units on the right side with minimum Hc (Units 0 and 16) and maximum B_s (Units 4 and 6).

Recommendation for optimum properties:

- Iron: Medium to high iron (Figure 6.25)
- Boron: Medium to low B (Figure 6.26)
- Copper: Medium to high Cu (Figure 6.26)
- Silicon: Medium to high Si for low H_c; low Si for high B_s (Figure 6.27)

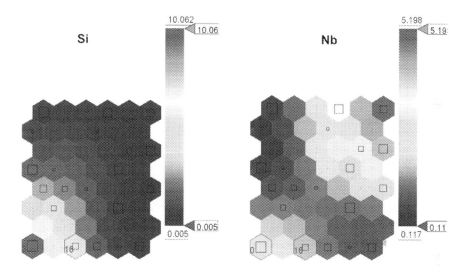

FIGURE 6.27   Distribution of silicon and niobium in the 2-D space.

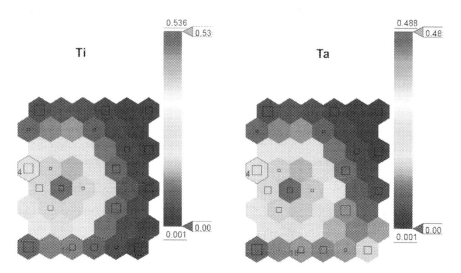

FIGURE 6.28   Distribution of titanium and tantalum in the 2-D space.

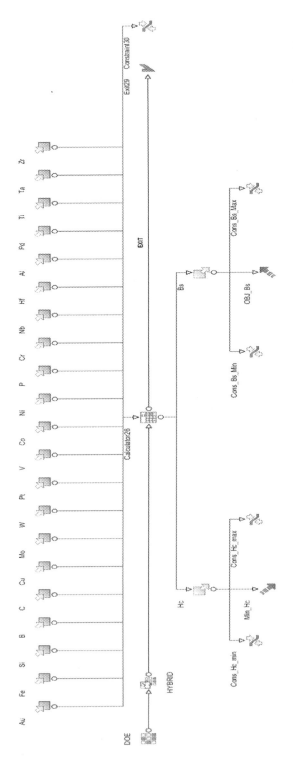

FIGURE 6.29 Workflow developed in modeFRONTIER for determining new chemistries of new magnets with maximum B_s and minimum H_c.

- Niobium: Medium Nb for low H_c; low Nb for high B_s (Figure 6.27)
- Titanium: Low Ti for low H_c; medium to low Ti for high B_s (Figure 6.28)
- Tantalum: Low Ta for low H_c; medium to low Ta for high B_s (Figure 6.28)

From the previous analysis, we were able to discover a few features through SOM analysis from a limited data set of 83 candidates:

1. Elements and properties are clustered together in a component chart (Figure 6.23). Such correlations have been observed in open literature on soft magnets.
2. Rather than focusing on the entire data set, it is beneficial to focus on a group of candidates for better understanding of the correlations between the alloying elements and between the alloying elements and targeted properties.
3. Element distribution chart gives an idea regarding the composition bounds for which one can expect to achieve optimum properties for design and discovery of new soft magnets.

6.6.2 SUPERVISED MACHINE LEARNING AND MULTI-OBJECTIVE OPTIMIZATION

In this work, we used a set of AI-based algorithms for developing surrogate models for B_s and H_c. Multi-objective optimization problem was formulated and the new alloys are expected to possess low coercivity (H_c) and high saturation magnetization (B_s); thus, maximize B_s and minimize H_c. We used modeFRONTIER software, and the workflow is shown in Figure 6.29.

6.7   CONCLUSIONS

This case study covers several aspects of materials design. A user can apply this strategy to any other alloy system. The main aspect of this chapter is calibration of CALPHAD-based models. In the beginning, we calibrated these models for simulating precipitation kinetics of magnetic phase from an amorphous precursor. Thereafter, we utilized these models for various time scales. We demonstrated ways of calibrating these models by which a user can get the desired result.

Application of AI-based algorithm is another aspect of this work. All the data generated from the experiments and simulations can be efficiently utilized by following the physics of the problem. Data from the literature can be the starting point, and a user can build their approach on the basis of data analysis performed on the data reported in literature.

The first step is to demonstrate improvements with the existing parameters and perform a few experiments. Finally, one can attempt discovery of novel composition for a given alloy, or completely discover a new alloy.

7 Case Study #7: Nickel-Based Superalloys: Combined Machine Learning-CALPHAD Approach

7.1 INTRODUCTION

Ni-based superalloy has been discussed in Chapter 2. The Ni-based superalloy system under study contains 20 alloying elements. This work used the data that has been used in our previous publications. There are 120 tested alloys in the initial database. All of these alloys are tested for estimating stress to rupture (N/mm^2) and the time to rupture (h) values. Response surfaces are generated for two properties: stress to rupture and time to rupture, followed by multi-objective optimization of these properties. Design optimization was performed in four cycles to check on improvements over the previous cycles. Each iteration yielded 20 new alloys. Out of these 20 alloys, there are a few candidates whose properties are better than the candidates included in the training set. At the same time, a few candidates' properties were similar to the alloys included in the initial training set.

In this work, we use the entire data set of 200 alloys. This database is examined under the framework of the CALPHAD approach to understand the difference between the experimentally verified Pareto-optimized predictions in each cycle, the initial training set of 120 alloys, and the IOSO predictions whose properties are similar to the properties of initial 120 alloys. We apply machine learning algorithms for finding meaningful correlations, visualization, and inverse determination of chemistry/composition of alloy for targeted properties.

7.2 CALPHAD APPROACH: THERMODYNAMICS/PHASE TRANSFORMATION

In one of our previous work, we utilized the same data set and performed multi-objective design optimization through a set of optimization algorithm (Jha et al. 2014). Figure 7.1 shows the optimized results (Pareto-front) obtained through that analysis (Jha et al. 2015). In Figure 7.1, one can observe that the Pareto-front obtained through each of the approches contain hundreds of canditate alloys. A few candidates were selected for further analysis under the framework of CALPHAD approach (Figure 7.2). A commercial software, FACTSAGE was used for this

DOI: 10.1201/9781003167372-7

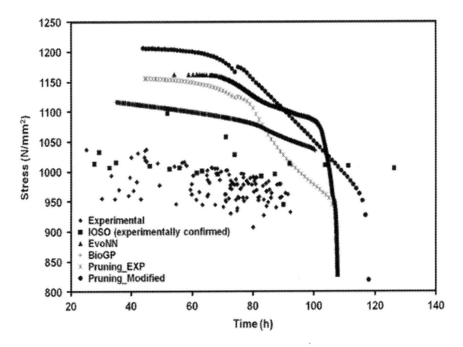

FIGURE 7.1 Comparison between Pareto-frontiers obtained through various AI-based approaches. Experimentally verified initial 120 alloys, and IOSO-software based predictions are plotted for comparison (Jha et al. 2015).

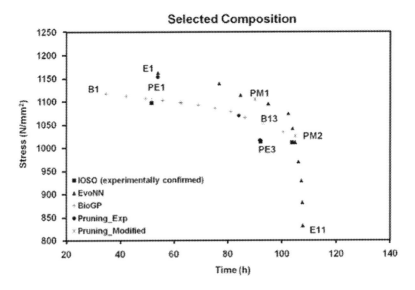

FIGURE 7.2 Candidate alloys selected for thermodynamic analysis through FACTSAGE software (Jha et al. 2015).

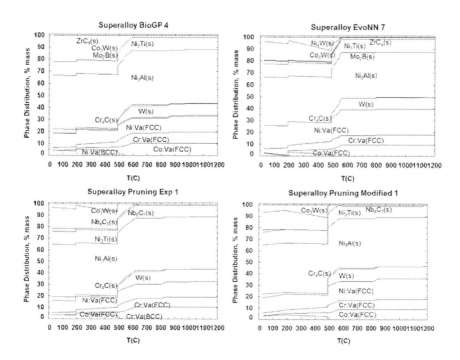

FIGURE 7.3 Phase stability analysis for a few candidate alloys that are part of the Pareto-frontier in Figure 7.1 and Figure 7.2 (Jha et al. 2015).

analysis. These candiadte alloys were chosen on the basis of their distance from experimentally-verified IOSO software predictions. Figure 7.3 shows the comparison between stability of various phases obtained through FACTSAGE software.

In this work, we use commercial software, Thermocalc and TCNI8 database (Ni-base superalloy database), to perform phase stability calculations. Equilibrium calculations are performed at 975°C, as time-to-rupture calculations are performed at this temperature. For the given system, Ni, C, Cr, Co, W, Mo, Al, Ti, B, Nb, Ce, Zr, and Y are identified to affect the bulk properties, while elements such as S, P, Fe, Mn, Si, Pb, and Bi are treated as extraneous impurities. Concentrations of Nb, B, Ce, Zr, and Y are kept constant in all of the alloys at 1.1%, 0.025%, 0.015%, 0.04%, and 0.01%, respectively. Variable bounds of the rest of the elements, namely C, Cr, Co, W, Mo, Al, and Ti have been tabulated in Table 7.1.

Cerium is not included in the TCNI8 database and, since it is fixed at a low amount of 0.015 weight %, we exclude it from our calculations and the balance is added to nickel.

One peculiar thing we noticed is that FCC_L12 exists in the TCNI8 database in four forms, where one of them is ordered (FCC_L12#3) and the other three are disordered (FCC_L12#1, FCC_L12#2, and FCC_L12#4). Both DIS_FCC_A1#2 and FCC_L12#1 can be used as a matrix phase during the precipitation simulation as both are disordered. Critical phases that exist in Ni-based superalloys for our work, the way the phases are notified in TCNI8 database in Thermocalc, along with their lattice occupancy, are tabulated in Table 7.2.

TABLE 7.1

Variables Bounds (Concentrations in wt. %) of 7 Major Alloying Elements

	C	Cr	Co	W	Mo	Al	Ti
Minimum	0.13	8.0	9.0	9.5	1.2	5.1	2.0
Maximum	0.20	9.5	10.5	11.0	2.4	6.0	2.9

TABLE 7.2

Critical Phases in Ni-based Superalloy, TCNI8 Notation and Their Lattice Occupancy

#	Critical Phases		Phase in TCNI8 Database		Lattice Occupancy
1.	Matrix	γ	DIS_FCC_A1#2		$(Al,Co,Cr,Mo,Nb,Ni,Ti,W,Y,Zr)_1$ $(B,C,VA)_1$
2.			FCC_L12#1 Disordered		$(Al,Co,Cr,Mo,Nb,Ni,Ti,W,Y,Zr)_{0.75}$ $(Al,Co,Cr,Mo,Nb,Ni,Ti,W,Y,Zr)_{0.25}$ $(B,C,VA)_1$
3.	GCP	γ'	FCC_L12	#2 Disordered	$(Al,Co,Cr,Mo,Nb,Ni,Ti,W,Y,Zr)_{0.75}$
4.				#3 Ordered	$(Al,Co,Cr,Mo,Nb,Ni,Ti,W,Y,Zr)_{0.25}$
				#4 Disordered	$(B,C,VA)_1$
5.	Carbides		M6C		$(Co,Ni)_2(Mo,Nb,W)_2$ $(Co,Cr,Mo,Nb,Ni,W)_2\ (C)_1$
6.			M23C6		$(Co,Cr,Ni)_{20}\ (Co,Cr,Mo,Ni,W)_3\ (C)_6$
7.	TCP	μ	MU_PHASE	#1 Ordered	$(Al,Co,Cr,Mo,Nb,Ni,W)_1$
8.				#2 Ordered	$(Al,Co,Cr,Mo,Nb,Ni,W)_2$
9.				#3 Ordered	$(Al,Co,Cr,Mo,Nb,Ni,W)_6$ $(Al,Co,Cr,Mo,Nb,Ni,W)_4$
10.		σ	SIGMA Ordered		$(Al,Co,Cr,Mo,Nb,Ni,Ti,W)_{10}$ $(Al,Co,Cr,Mo,Nb,Ni,Ti,W)_4$ $(Al,Co,Cr,Mo,Nb,Ni,Ti,W)_{16}$
11.		τ	TAU		$(Co,Ni)_{20}\ (B)_6\ (B,VA)_6$ $(Al,Cr,Mo,Ti,W,Zr)_3$
12.		R	R_PHASE		$(Co,Cr,Ni)_{27}\ (Mo,W)_{14}$ $(Co,Cr,Mo,Ni,W)_{12}$
13.	Other		M2B_TETR		$(Al,Co,Cr,Mo,Nb,Ni,W)_2\ (B)_1$
14.	phases		NI17Y2		$(Al,Ni)_1\ (Y)_{0.12}$

The phase transformation diagrams obtained through Thermo-Calc software for Alloy #31, Alloy #183, and Alloy #193, has been included in the supplementary documents.

Alloy #31 is part of the initial data set of 120 experimentally verified alloys. Based on these 120 alloys, a multi-objective design optimization is formulated, and

20 alloys were developed in each cycle. In each cycle of 20 alloys, few alloys outperform the alloys that are part of the training set. These alloys are placed into a group for each cycle and have been mentioned in this chapter as Pareto-optimized alloys. The rest of the alloys in each cycle perform similar to the alloys in the training set, and these alloys have been referred to as the "IOSO optimized non-Pareto" set in this chapter.

7.3 APPLICATION OF MACHINE LEARNING ALGORITHMS

7.3.1 Visualization and Correlation

We use the concept of parallel coordinate chart as a visualization tool to draw meaningful correlations. Thereafter, we estimate correlations between the desired phase and alloying elements, desired phase and properties of interest, and critical strengthening phase γ' and detrimental TCP μ phase.

Figure 7.4 shows the parallel coordinate chart (PCC) representation of critical phases for alloy #31, 183, 188, and 193. Alloy 183 belongs to the group/class of Pareto-optimized IOSO prediction.

From Figure 7.3, and 7.4, we can observe some differences with respect to amount of phases estimated through CALPHAD-based software: FACTSAGE and Thermocalc, respectively. But, from these figures, we cannot draw any meaningful conclusion, even though these alloys have different properties.

Figure 7.5 shows the correlation matrix for the Pearson's correlation coefficient between the alloying elements and the critical phases for the alloys that were part of the Pareto-optimized IOSO predictions. Unique observations are listed in the following text:

1. FCC_A1#2: Even though correlation coefficients are low, an increase in C, Cr, and Co will lead to an increase in FCC_A1#2 to a certain extent, while Al seems to have a negative impact on FCC_A1#2. For Ni, Ti, W, and Mo, the coefficient is too low to draw any meaningful conclusion.
2. FCC_L12#1: Increase in Al will lead to a decrease in FCC_L12#1 as the correlation coefficient if quite high to be ignored. Ni seems to stabilize FCC_L12#1 to a certain extent, whereas all the other elements seem to affect FCC_L12#1 negatively.
3. FCC_L12#3: Increase in Co and Cr will lead to increase in FCC_L12#3, whereas Ti negatively affects FCC_L12#3 to a certain extent. For other elements, the correlation coefficient is low.
4. MU_PHASE#1: Increase in Mo will lead to an increase in MU_PHASE#1 as the correlation coefficient is quite high. Apart from Ni, all other elements seem to stabilize MU_PHASE#1.
5. MU_PHASE#2: Increase in Al will lead to an increase in MU_PHASE#2 as the correlation coefficient is quite high. Apart from Ni, all other elements seem to stabilize MU_PHASE#2.
6. MU_PHASE#3: Increase in Al and Ti will lead to an increase in MU_PHASE#3 to a certain extent. Apart from Ni and Co, other elements

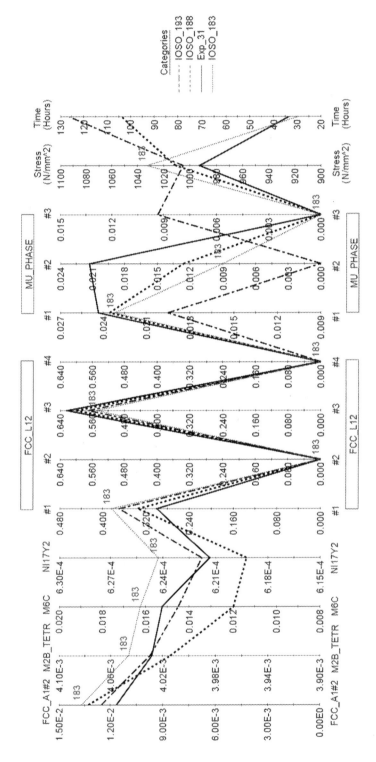

FIGURE 7.4 Parallel coordinate chart for alloys #31, 183, 188, and 193.

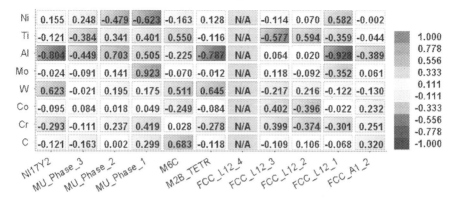

FIGURE 7.5 Correlation matrix: Alloying elements and the critical phases for alloys Pareto-optimized IOSO predictions.

have a negative impact on MU_PHASE#3 even though the correlation coefficient if low.

7. M6C: Ti and W are known carbide formers and from the correlation charts it seems to have a positive impact on the amount of M6C. The correlation coefficient is lower for other elements.

Moving forward, we plotted the correlation matrix for critical phases and stress and time to rupture for Pareto-optimized IOSO predictions. Figure 7.6 shows the correlation matrix for Pearson's correlation coefficient between the critical phases and stress and time to rupture for the alloys that were part of the Pareto-optimized IOSO predictions. Unique observations are listed in the following text:

1. Stress and time are conflicting in nature as the correlation coefficient is quite high.
2. Even though correlation coefficients are quite low:
   a. Stress to rupture: Increase in FCC_A1#2, FCC_L12#3, MU_PHASE#1, and MU_PHASE#2 will lead to an increase in stress to rupture, whereas, an increase in FCC_L12#1 and MU_PHASE#3 will lead to a decrease in stress to rupture.
   b. Time to rupture: Increase in FCC_L12#1 and MU_PHASE#3 will lead to an increase in time to rupture, whereas an increase in FCC_A1#2, FCC_L12#3, MU_PHASE#1, and MU_PHASE#2 will lead to a decrease in time to rupture.
3. Increase in FCC_A1#2 will lead to an increase in both FCC_L12#1 and FCC_L12#3 and an increase in MU_PHASE#3, whereas it will lead to a decrease in MU_PHASE#1 and MU_PHASE#2.
4. Increase in FCC_L12#1 will lead to a decrease in MU_PHASE#1 and MU_PHASE#2, while it will lead to an increase in MU_PHASE#3.
5. Increase in FCC_L12#3 will lead to a decrease in MU_PHASE#2, while it will lead to an increase in MU_PHASE#1 and MU_PHASE#3.

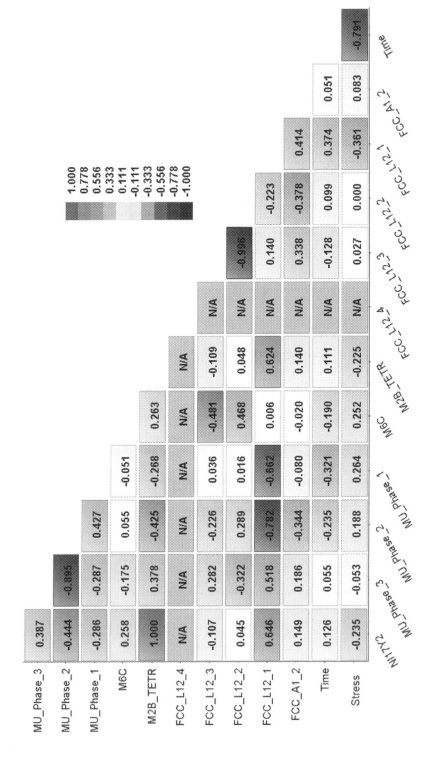

FIGURE 7.6 Correlation matrix: Stress, time, and the critical phases for alloys Pareto-optimized IOSO predictions.

7.3.2 INVERSE DESIGN OF CHEMISTRY/COMPOSITION FOR GIVEN STRESS AND TIME TO RUPTURE

We used Citrine Informatics to develop metamodels for stress and time to rupture as a function of the critical phase (O'Mara, Meredig, and Michel, 2016). Citrine informatics uses "random forest" algorithm for developing meta-models. It was followed by inverse determination of the amount of critical phases responsible for the given stress and time to rupture. Thereafter, we developed meta-models for chemistry of the alloy as a function of critical phases. It was followed by inverse determination of the chemistry/composition of the alloy that will yield the critical phases predicted previously.

The chemistry/composition predicted by machine learning (CITRINE) was analyzed through the CALPHAD by performing equilibrium calculations at 975°C through the CALPHAD (Thermocalc) approach. The amount of critical phases predicted through machine learning (CITRINE) and the CALPHAD (Thermocalc) approach are in close proximity. Thus, we were able to inversely determine the chemistry/composition of a class of nickel-based superalloys for a given stress and time-to-rupture value.

7.4 INVERSE DETERMINATION OF CHEMISTRY OF ALLOY FOR A GIVEN VALUE OF STRESS AND TIME TO RUPTURE

In this work, we use an open-source platform, Citrine Informatics (O'Mara, Meredig, and Michel, 2016), to develop meta-models and perform an inverse determination of critical phases and chemistry of the alloy. We use the candidates that are part of the IOSO-optimized prediction. The variables and properties were scaled between 0 and 1.

Work flow for Citrine informatics:

1. First, import the data and define the properties of interest as PROPERTY and variables/parameters affecting the properties as CONDITION.
2. After successful import:
   a. Refer to DATAVIEW tab
   b. Select CREATE NEW DATAVIEW and give it a name
   c. Check if the PROPERTY and CONDITION are labeled correctly
   d. If needed, select the minimum and maximum values for CONDITION and PROPERTY
   e. Press SAVE and return to the MATRIX tab.
3. Citrine informatics uses the random forest algorithm to develop meta-models for PROPERTY as a function of CONDITION. Depending on the size of the data set, it takes a few minutes to develop the model.
4. Check the model for finding correlations. We already reported correlations, so we skipped this part.
5. Refer to DESIGN tab:
   a. At the top of the window, the user must enter a number in minutes for "design duration". Longer duration is helpful for achieving the best material that will appear after we execute this step.

b. Next comes, number of candidates that a user needs. Here the user needs to enter a value.
c. For input constraints, either provide a fixed value or provide minimum and maximum values for the variables mentioned as CONTITION.
d. For target constraints, provide minimum and maximum values for the targeted property mentioned as PROPERTY.
e. Additionally, there is one target option, where a user can MAXIMIZE or MINIMIZE a target (PROPERTY). If a user doesn't want to optimize any target (PROPERTY), they can select "none".
6. After execution, candidates are ranked and can be differentiated by term "Citrine Score".

In this work, we scaled the entire data set between 0 and 1. We divided our problem in three parts:

1. Developed meta-models for targeted properties (assigned as PROPERTY): Stress and time to rupture; as a function of critical phases (assigned as CONDITION): DIS_FCC_A#2, FCC_L12#1, FCC_L12#3, MU_PHASE#1, MU_PHASE#1 and MU_PHASE#3.
   a. Design Duration = 30 minutes and number of candidates = 5.
   b. Input Constraints: Minimum = 0 and Maximum = 1.
   c. Target Constraints: Minimum = 0 and Maximum = 1.
   d. Target: Maximize stress to rupture.
   e. Due to software limitations and scarce data set, software gave us a few candidates with identical stress and time value between 0 and 1. The values were rescaled as Stress = 1,048 N/mm^2 and Time to rupture = 62 hours.
   f. For these given stress and time, we inversely determined the amount of critical phases. Rescaled values have been tabulated in Tables III and IV.
   g. We selected two candidates and move to step 2.
2. Developed meta-models for composition of alloy (assigned as PROPERTY) C, Cr, Co, W, Ti, Mo, and Al; as a function of critical phases (assigned as CONDITION): DIS_FCC_A#2, FCC_L12#1, FCC_L12#2, FCC_L12#3, MU_PHASE#1, MU_PHASE#1, and MU_PHASE#3.
   a. Design Duration = 30 minutes and Number of Candidates = 5.
   b. Input constraints are fixed at values of critical phases obtained from step 1, for which the rescaled values have been tabulated in Tables 7.4 and 7.5.
   c. Target Constraints: Minimum = 0 and Maximum = 1.
   d. Target: None.
   e. Software gave ONLY one candidate for either case (for 2 candidates selected in step 1).
   f. Rescaled value of composition is tabulated in Table 7.3 and for critical phases in Tables 7.4 and 7.5.

TABLE 7.3

Composition of Alloys Determined through CITRINE

	Co	Mo	Al	Cr	C	Ti	W
Alloy 1	10.0734	1.6763	5.6355	9.1052	0.1876	2.6639	10.4167
Alloy 2	10.0770	1.6648	5.6347	9.1020	0.1873	2.6623	10.4194

TABLE 7.4

Comparison between Phases Predicted through Machine Learning (CITRINE) and through CALPHAD (Thermocalc) for Alloy 1

Alloy 1	Machine Learning (CITRINE)	CALPHAD (Thermocalc)	Error (%)
DIS_FCC_A#2	0.0135	0.0134	1.3551
FCC_L12#1	0.3390	0.3391	0.0246
FCC_L12#2	0.0109	0.0000	NA
FCC_L12#3	0.5866	0.5924	0.9937
MU_PHASE#1	0.0232	0.0233	0.2916
MU_PHASE#2	0.0122	0.0123	0.7925
MU_PHASE#3	0.0012	0.0000	NA

TABLE 7.5

Comparison between Phases Predicted through Machine learning (CITRINE) and through CALPHAD (Thermocalc) for Alloy 2

Alloy 2	Machine Learning (CITRINE)	CALPHAD (Thermocalc)	Error (%)
DIS_FCC_A#2	0.0135	0.0133	1.2715
FCC_L12#1	0.3369	0.3396	0.7873
FCC_L12#2	0.0000	0.0000	0.0000
FCC_L12#3	0.5864	0.5922	0.9875
MU_PHASE#1	0.0231	0.0231	0.1123
MU_PHASE#2	0.0130	0.0123	5.5719
MU_PHASE#3	0.0004	0.0000	NA

3. Compare Machine Learning (CITRINE) and through CALPHAD (Thermocalc) approach.
   a. Used the composition reported in Table 7.3 and performed equilibrium calculations at 975°C in Thermocalc.
   b. Checked the difference between amount of critical phases determined through inverse design: Machine learning (CITRINE) and that calculated through CALPHAD (Thermocalc).
   c. Tabulated the comparison along with the error in Tables 7.4 and 7.5.

Table 7.3 shows the chemistry determined through CITRINE in step 2. The amount of critical phases from step 1 were used to determine the chemistry. The amount of critical phases for both the alloy have been tabulated in Tables 7.4 and 7.5.

Tables 7.4 and 7.5 show the comparison between the amounts of phases determined through machine learning (CITRINE) and through the CALPHAD (Thermocalc) approach for alloy 1 and alloy 2 mentioned in Table 7.3.

For alloy 1 (Table 7.3), critical phases are tabulated in Table 7.4. From Table 7.4, we can observe that the difference between the phases predicted through machine learning and calculated through the CALPHAD approach is about 1%. Through a machine learning approach, we observed traces of FCC_L12#2 and MU_PHASE#3, but these phases were not observed through the CALPHAD approach.

For alloy 2 (Table 7.3), critical phases are tabulated in Table 7.5. From Table 7.5, we can observe that the difference between the phases predicted through machine learning and calculated through the CALPHAD approach is about 1% for all the phases except MU_PHASE#2 (5.6%). For alloy 2, no FCC_L12#2 was observed through both machine learning and the CALPHAD approach. Through the machine learning approach, we observed traces of the MU_PHASE#3, but it was not observed through the CALPHAD approach.

For alloy 2 (Tables 7.3 and 7.5), FCC_L12#2 is predicted through machine learning (CITRINE) came to be 0 and FCC_L12#2 was absent in equilibrium calculations through CALPHAD (Thermocalc). Thus, from Table 7.5, we can confirm that it is possible to work with missing data through the machine learning process as we could predict FCC_L12#2 accurately and confirmed our findings through the CALPHAD approach. Looking at the complexity of a problem, error (%) is low at about 1% for most of the cases. Thus, improvements in meta-model development will further help us in understanding the complex problem of finding correlations between chemistry, critical phases, and desired property for nickel-based superalloys.

7.5 CONCLUSION FROM INVERSE DESIGN

In this work, we presented a case study for inverse design. Experimentally verified multi-component nickel-based superalloys are analyzed under the framework of CALPHAD and the machine learning approach. Ni-based superalloys' composition is inversely determined for a set of given properties.

Unique findings are as follows:

1. Introduced a novel approach of bridging the gap between alloy composition and targeted properties by efficiently utilizing CALPHAD and machine learning approach.
2. Determined correlations between:
   - Alloying element and critical phases.
   - Critical phases and targeted properties: Stress and time to rupture.
   - Strengthening gamma prime (γ') phases and detrimental TCP (μ) phases.

3. Inverse determination of:
   - Critical phases for a given targeted property.
   - Chemistry/composition from amount of critical phases predicted in the previous step.
4. The amount of critical phases predicted through machine learning approach and calculated from the CALPHAD approach (for the composition predicted from machine learning) are in close proximity.

This approach can be used for designing other systems of alloys too.

8 Case Study #8: Design of Aluminum Alloys: Combined Machine Learning-CALPHAD Approach

8.1 INTRODUCTION: BACKGROUND

In this chapter, we present a case study on designing new chemistries of heat-treatable aluminum alloys (2XXX, 6XXX, and 7XXX). Additionally, we have explored the scope of addition of scandium in these alloys. The scandium addition improves multiple properties of aluminum-based alloys, but it is an expensive process. Scandium is added in small amounts, mostly within 1–2 weight %.

In this chapter, we have presented solidification and heat-treatment simulations under the framework of the CALPHAD approach for studying the precipitation kinetics of the Al_3Sc phase. Three compositions were chosen for this simulation for the 6XXX class of alloys. These compositions were obtained through a combined CALPHAD-machine learning approach, where three classes of heat-treatable aluminum-based alloys were analyzed (Jha and Dulikravich 2020, 2021).

In a previous work (Jha and Dulikravich 2020, 21), we performed a computational study of the phase stability of precipitation strengthening phases in 2XXX, 6XXX, and 7XXX series aluminum-based alloys that included scandium. Phase stability data were analyzed through the deep learning artificial neural network (DL-ANN) models, where code was developed in Python using Tensorflow (Abadi et al. 2016) and Keras (Chollet 2015, Keras 2020) libraries. Data obtained through the DL-ANN model was further analyzed through the concept of self-organizing maps. SOM analysis provided us with novel compositions for 2XXX, 6XXX, and 7XXX alloys. In this chapter, we have used three (3) of those predictions for the 6XXX alloy for performing solidification and heat-treatment simulations. Initially, the Scheil-Gulliver solidification simulation was carried out to determine the formation of the Al_3Sc phase from the melt. Subsequently, a heat treatment simulation was performed to investigate the kinetics of the precipitation (nucleation, growth, and coarsening) of the Al_3Sc phase in these alloy classes.

DOI: 10.1201/9781003167372-8

8.2 MATERIALS AND METHODS: CALPHAD APPROACH: THERMODYNAMICS/PHASE TRANSFORMATION AND PRECIPITATION KINETICS

In this study, the aluminum alloy classes 2XXX, 6XXX, and 7XXX are selected because these alloy classes are heat-treatable (De Luca et al. 2018, Røyset and Ryum 2005). These variable limits for the chemical composition of the 2XXX, 6XXX, and 7XXX alloy series are summarized in Table 8.1. It should be noted that in 2XXX a total of 12 alloy elements are considered. In the 7XXX series, we thought about adding Zr with Sc while dropping V, while for the 6XXX series we did not include both Zr and V. Equilibrium calculations were undertaken to stabilize the metastable phases of these alloys.

8.2.1 CALPHAD: IDENTIFICATION OF STABLE AND METASTABLE PHASES

The precipitation sequence of critical phases in aluminum-based alloys are as follows:

- 2XXX: *Supersaturated solid solution* →*GP-zones* → θ'' → θ *(Al$_2$Cu) phase.*
  θ (Al$_2$Cu) phase is a stable phase. 2XXX-type alloys achieve superior strength when θ'' and θ' phases are the predominant phases (Røyset and Ryum 2005).
- 6XXX: *Supersaturated solid solution* →*GP-zones* → β'' → β *(Mg$_2$Si).*
  β (Mg$_2$Si) phase is a stable phase, while β'' is the predominant phase. β'' is a precipitation hardening phase and it is observed after aging (Røyset and Ryum 2005).
- 7XXX: *Supersaturated solid solution* →*GP-zones* → η' → η *(Mg$_2$Zn).*

TABLE 8.1

Minimum and Maximum Concentrations for Each of the 12 Alloying Elements (wt. %) for Three Series of Al-Sc-based Alloys (Jha and Dulikravich 2021)

Element	2XXX Series		6XXX Series		7XXX Series	
	Min.	Max.	Min.	Max.	Min.	Max.
Si	0.20	1.20	0.20	1.80	0.12	0.50
Fe	0.30	0.50	0.10	0.70	0.15	0.50
Cu	3.80	6.80	0.10	0.40	0.10	2.40
Mn	0.20	1.20	0.05	1.10	0.05	0.70
Mg	0.02	1.80	0.35	1.40	0.80	3.70
Cr	0.00	0.10	0.00	0.35	0.00	0.30
Zn	0.10	0.25	0.05	0.25	3.80	8.30
Ti	0.02	0.15	0.00	0.20	0.01	0.20
V	0.00	0.15	0.00	0.00	0.00	0.00
Zr	0.00	0.25	0.00	0.00	0.00	0.20
Sc	0.00	10.00	0.00	10.00	0.00	10.00
Al	Balance to 100.00		Balance to 100.00		Balance to 100.00	

η (Mg$_2$Zn) phase is a stable phase. This class of alloys achieves superior strength when η' and η are the predominant phases. Maximum hardness is achieved when η' is the predominant phase (Røyset and Ryum 2005).

Assadiki et al. (2018) have provided important guidelines that are helpful for studying various stable and metastable phases alloys within the framework of the CALPHAD approach for 2XXX and 6XXX. In this work, we used the thermodynamic database TCAL5 (Thermo-Calc Software TCAL5 2018) and the mobility database MOBAL4 (Thermo-Calc Software MOBAL4 2018) in a commercial software Thermocalc 2018B (Andersson et al. 2002) for studying the phase stability of various phases in 2XXX, 6XXX, and 7XXX alloys. As mentioned previously, in 2XXX, the oversaturated solid solution transforms into GP zones that transform into θ'', followed by θ', which eventually transforms into a stable θ (Al2Cu) phase. Reported literature on various precipitates in aluminum alloys point out that θ'' closely resembles GPII zones in 2XXX, while θ-Al$_2$Cu is the stable phase (Andersen et al. 2018, Thermocalc TCAL5 2020). In 6XXX, β-Mg$_2$Si is the stable phase and β'' resembles GPII phases or GPII zones (Andersen et al. 2018, Thermocalc TCAL5 2020). In 7XXX alloys, η-MgZn$_2$ is the stable phase and η' is the metastable phase, also known as GP zones (Kumar and Padture 2018). η-MgZn$_2$ phase has a C14 structure. For 7XXX alloys, V_PHASE is usually considered an η-MgZn$_2$ phase. But in the TCAL5 database, C14_LAVES_PHASE also has a C14 structure (Thermocalc TCAL5 2020). Thus, we performed calculations for both V_PHASE and C14_LAVES_PHASE, as the structure of GP zones in 7XXX alloys is a matter of debate.

In the TCAL5 database (Thermocalc TCAL5 2020), AL3X can be considered as the notation for the Al$_3$Sc phase. The component X in AL3X can have a different meaning. X can be a combination of Sc and Zr, and in most cases traces of Mg are also observed. In this work, we performed the Scheil-Gulliver solidification simulation to estimate the temperature where AL3X completely transforms into the Al$_3$Sc phase. The Scheil-Gulliver solidification simulation also provided liquidus temperature and the amount of the Al$_3$Sc phase in the melt at the end of the solidification. The Al$_3$Sc phase must be present in the melt at the end of the solidification process. The presence of the Al$_3$Sc phase in the melt is helpful for nucleation and growth of Al$_3$Sc grains during the aging process.

The CALPHAD approach uses Gibbs-energy minimization as a criterion for determining the formation and stability of a given phase (Andersson et al. 2002, Assadiki et al. 2018). One may not observe any metastable phase for 2XXX, 6XXX, and 7XXX alloys while performing equilibrium calculations, although superior properties are achieved in these alloys as a result of the optimum amount of stable and metastable phases (Røyset and Ryum 2005). That is, experimentally, stable and metastable phases exist in the final microstructure, but may not coexist in phase transformation calculation in Thermocalc (Thermocalc TCAL5 2020). A metastable phase can be preferentially stabilized within the framework of the CALPHAD approach by suppressing the most stable phase for that system while performing phase stability calculations. In this case study, as well as other case studies discussed in this book, this strategy was followed for stabilizing various stable and metastable phases in aluminum, titanium, titanium-based biomaterials, nickel-based superalloys, and soft magnetic FINEMET-type alloys. In this case study, data were generated for the Al_3Sc phase for about 1,200 sets of composition and temperature. Thereafter, data for various stable and metastable phases for 2XXX, 6XXX, and 7XXX alloys were generated for these sets of composition and temperature.

8.3 RESULTS: APPLICATION OF MACHINE LEARNING ALGORITHMS

8.3.1 DEEP LEARNING ARTIFICIAL NEURAL NETWORK (DL-ANN) MODEL

Based on literature (Andersen et al. 2018, Assadiki et al. 2018, Haidemenopoulos et al. 2010) and our developed DL-ANN model (Jha and Dulikravich 2020), performance metrics for models for various phases in 2XXX, 6XXX, and 7XXX alloys have been tabulated in Table 8.2.

The architecture of the DL-ANN model chosen in this work has been tabulated in Table 8.2. The mean absolute error (MAE) in Table 8.2 seems to be large at first glance, but all of these values will be rescaled. The amount of phases calculated from the CALPHAD approach are between 0 and 0.1. The MAE values of 0.01 will become 0.001 after re-scaling, and the maximum MAE of 0.047 will become

---

TABLE 8.2

Performance Metrics for DL-ANN Models over Testing/Validation Set in 2XXX,6XXX, and 7XXX Alloys (Jha and Dulikravich 2021)

Alloy	Phase	DL-ANN Architecture	Error Metrics (Testing Set)	
			Mean Square Error (MSE)	Mean Absolute Error (MAE)
2XXX	THETA_PRIME (θ')	90-180-180-180	4.47e-4	0.01535
	THETA_DPRIME (θ'')	60-120-120-120	9.87e-4	0.01972
	AL2CU_C16 (θ)	80-160-160-160	0.01086	0.04221
	S_PHASE	50-100-100-100	4.24E-4	0.01077
	AL3X (Al_3Sc)	80-160-160-160	6.69e-4	0.0207
6XXX	MG2SI_C1(β)	70-140-140-140	6.63e-4	0.01911
	BETA_PRIME (β')	80-160-160-160	8.32e-4	0.0175
	BETA_DPRIME (β'')	80-160-160-160	1.79e-3	0.01075
	AL3X (Al_3Sc)	80-160-160-160	6.09e-4	0.01898
7XXX	C14_LAVES (η')	50-100-100-100	6.63e-4	0.01892
	T_PHASE	100-200-200-200	1.56e-3	0.01877
	V_PHASE (η-$MgZn_2$)	80-160-160-160	8.31e-3	0.04712
	S_PHASE	50-100-100-100	1.76e-3	0.01604
	AL3X (Al_3Sc)	60-120-120-120	2.13e-5	2.9961e-3

0.0047. The maximum amount of the phase is about 0.1, so, the MAE will be between 1% and 5% of the maximum amount. For a complex problem like this, 11–13 design parameters (composition and temperature), a large database and several missing values, and an error between 1% and 5% are acceptable. There is always room for improvement, but ANN models are prone to overfitting. Therefore, care must be taken when choosing a model and its use as a predictive tool. Throughout this book, priority is always given to the physics of the problem, or the metallurgical concepts and governing equations. In this case study, information from the physical metallurgy of aluminum alloys is the reference for selecting DL-ANN models, while using statistical and artificial intelligence concepts as guides.

DL-ANN models have a different structure. In a model of structure of 50-100-100-100, the numerical value of 50 represents the number of neurons in the first hidden layer of DL-ANN. Each DL-ANN model has four hidden layers, and the three values of 100 represent the number of neurons in the other three hidden layers.

On the basis of our experience, we developed models with acceptable MSE and MAE performance measurements (calculated on the validation/test set). Sixty-seven percent of data was assigned to the training set, while 33% of data was assigned to the testing set. One can improve upon the model accuracy by adding more data in the training set, but care must be taken to avoid overfitting.

8.3.2 SELF-ORGANIZING MAP (SOM) ANALYSIS

Table 8.3 summarizes the performance metrics of SOM analysis for 2XXX, 6XXX, and 7XXX alloys. Table 8.3 shows that the topological error is low, whereas the quantization error is comparatively higher.

In this work, predictive models have already been developed by the TensorFlow/Keras libraries in Python (Abadi et al. 2016, Chollet 2015, Keras 2020) with enhanced predictive accuracy (Table 8.2). SOM will be used for understanding patterns in the data set, while for prediction and in selecting novel composition for 6XXX alloys in this case study.

8.3.3 6XXX ALLOYS

The SOM component plot for 6XXX alloys is been presented in Figure 8.1. MG2SI_C1 and BETA_PRIME are adjacent to each other. From the literature, we

TABLE 8.3

SOM Performance Metrics for 2XXX, 6XXX, and 7XXX Series Aluminum Alloys

	Quantization Error	Topological Error
2XXX	0.126	0.030
6XXX	0.109	0.034
7XXX	0.131	0.017

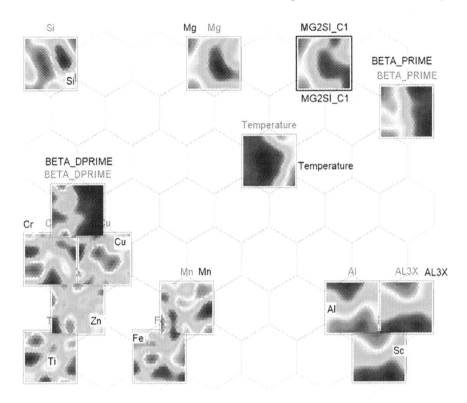

FIGURE 8.1 SOM plot of components for 6XXX series of Al-based alloys along with Sc and temperature.

know that MG2SI_C1 is stable, while BETA_PRIME is a metastable phase that eventually transforms into MG2SI_C1. The temperature is close to Mg, MG2SI_C1, and BETA_PRIME. Mg and MG2SI_C1 units are adjacent to each other, and their correlation can be understood through physical metallurgy. A SOM is able to detect correlations that can be verified from concepts of physical metallurgy. Even though the SOM approach is an unsupervised machine learning approach that doesn't work on principles of physical metallurgy, this demonstrates the efficacy of using the SOM algorithm in determining correlations among chemical composition and various stable and metastable phases. In the alloy data set, BETA_DPRIME is observed only for a few cases. On the SOM, it is clustered together with minor elements like Si, Zn, Mn, Fe, Cr, and Ti. AL3X, Sc, and Al are adjacent to each other in Figure 8.1 and from the literature this can be verified.

Figure 8.2 shows the distribution of precipitation hardening phases (mole fraction) for 6XXX alloys: BETA_PRIME, BETA_DPRIME, and MG2SI_C1 along with AL3X; Sc (weight %); and temperature (°C). SOMs consist of hexagonal cells. A user can look at the color bar adjacent to each subfigure to quantify the chemical composition and phases associated with a certain cell. The value associated with a cell on a SOM is the average value of all the candidates positioned at the vertices of a cell. The SOM looks pixelated, and hexagonal cells are not clearly visible in

FIGURE 8.2 SOMs showing a distribution of precipitation hardening phases (mole fraction) in 6XXX alloys: BETA_PRIME, BETA_DPRIME, and MG2SI_C1 along with AL3X, Sc (wt. %), and temperature (°C).

Figure 8.2 due to a large number of candidate alloys included in this study. SOM analysis can be performed for a small number of candidates (even 25–30), and for a large number of candidates due to its feature of preserving the topology of a data set. In Figure 8.2, the Sc and AL3X phases are correlated throughout the temperature regime. The BETA_PRIME phase transforms into BETA_DPRIME, which eventually transforms into MG2SI_C1 phases that are unstable at higher temperatures (Røyset and Ryum 2005). We have positioned SOMs of BETA_PRIME, BETA_DPRIME, and MG2SI_C1 together for visualization. One can observe that BETA_PRIME and BETA_DPRIME are absent at higher temperatures, while MG2SI_C1 phase is present in a small amount. BETA_DPRIME data from the CALPHAD analysis had lots of missing values when compared with data from other phases. Even with a small data set for BETA_DPRIME, SOMs were able to position their distribution in the same region of SOMs as BETA_PRIME. This information is important from a metallurgical point of view as a SOM algorithm is an unsupervised machine learning algorithm but still is able to catch information reported in literature on the physical metallurgy of aluminum. This demonstrates the efficacy of utilizing SOMs for pattern recognition in the alloy design domain of physical metallurgy.

The thermal treatment of aluminum alloys containing scandium is done in two stages. Annealing is performed around 100–200°C for precipitating BETA_PRIME, BETA_DPRIME, and MG2SI_C1 phases, while precipitating for the AL3X annealing temperature is above 300 °C.

Additionally, it is important that there is some amount of AL3X precipitated in the melt during solidifying stages.

The Sc addition in aluminum alloys is usually limited up to 2 weight % due to various complexities discussed in this chapter, our previous work, and the cited references. Since this is a book, we have taken the liberty to explore the effects of the Sc addition in amounts above 2 weight %. Based on the preceding computational effort, the three (3) most promising Al-Sc-based alloy from 6XXX class of aluminum alloys was chosen for performing solidification and precipitation kinetics simulations within the framework of the CALPHAD approach.

Candidates in Table 8.4 are analyzed for phase stability. The equilibrium amount of phases for these three candidates are tabulated in Table 8.5.

We performed solidification simulations to determine the temperature at which the AL3X precipitate will be fully Al_3Sc. Additionally, we estimate the amount (mole fraction) of Al_3Sc when AL3X is fully Al_3Sc, liquidus temperature, final amount of Al_3Sc, and the final temperature where solidification simulation stops. All the values are estimated, though the Scheil-Gulliver simulations are listed in Table 8.6. Liquidus temperature is the temperature above which all the alloy constituents are fully melted and represent a liquid. The Scheil-Gulliver solidification simulation for alloy 6XXX_3 is shown in Figure 8.3.

Figure 8.3 provides vital information on the precipitation sequence of different phases. The most important thing is to determine the temperature at which AL3X is 100% Al_3Sc. The Al_3Sc phase in the casting affects the quantity of the Al_3Sc phase obtained after heat treatment. It is advised that solidification must be carried out in a way that Al_3Sc is precipitated in the melt.

TABLE 8.4
Compositions of Three Candidates (6XXX) for Al-based Alloys Chosen for Solidification and Heat-treatment Simulation

Alloying Element	6XXX_1	6XXX_2	6XXX_3
Si	1.5497	1.51365	1.19879
Fe	0.57313	0.28235	0.29014
Cu	0.27893	0.23705	0.19349
Mn	0.63697	0.89761	0.56941
Mg	1.07076	0.95222	1.08212
Cr	0.1437	0.1189	0.14959
Zn	0.11882	0.18904	0.11832
Ti	0.13088	0.05924	0.13862
Sc	2.60077	5.2191	4.00005
Al	92.89525	90.53085	92.25947

TABLE 8.5
Equilibrium Number of Phases at Three Different Temperatures

Candidate:	Temperature	AL3X	BETA_PRIME
6XXX_1	163.70344 °C	0.04696	0.01786
6XXX_2	173.7694	0.10881	0.01581
6XXX_3	174.72572	0.09171	0.01792

TABLE 8.6
Critical Findings from Scheil-Gulliver's Solidification Simulation for 6XXX Alloys

	6XXX_1	6XXX_2	6XXX_3
Liquidus temperature (°C)	799.2	879.2	854.17
Temperature (fully Al_3Sc) (°C)	598.96	611.272	612.609
Al_3Sc (mole fraction)	0.05966	0.12339	0.09469
Solid (mole fraction)	0.8	0.74943	0.76729
Final temperature (°C)	505.263	506.352	531.242
Final solid (mole fraction)	0.9902	0.99048	0.99016
Final Al_3Sc (mole fraction)	0.05984	0.12393	0.09536

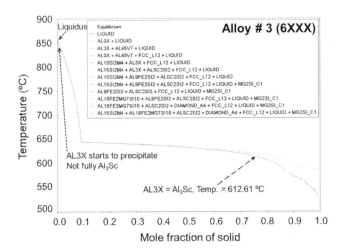

FIGURE 8.3   Scheil-Gulliver solidification simulation results for 6XXX_3.

From Table 8.6, one can observe that the liquidus temperature is affected by scandium. 6XXX_2 has the highest Sc content, and highest liquidus temperature. The amount of solid (mole fraction) when Al3X is fully Al₃Sc decreases with an increase in Sc. The amount of Al₃Sc will increase with Sc, but, alloy 6XXX_3 solidifies at a higher temperature when compared with the other two candidates. Finally, Al₃Sc is slightly higher than the Al₃Sc estimated at a temperature where AL3X fully transforms into Al₃Sc. This information will be helpful in designing heat treatment of the alloy.

Depending on the application, the Al₃Sc grain size can vary from 2 to 100 nm (Røyset and Ryum 2005). Strengthening in aluminum alloys is achieved by precipitation hardening and also mechanical working. Isothermal heat-treatment simulations are carried out between 300 and 450 °C, and the results are presented in Figure 8.4. One can observe that alloy 6XXX_2 has the maximum mean radius for all the temperatures. 6XXX_2 has the maximum Sc content, so it is expected. But, 6XXX_1 and 6XXX_3 have different Sc contents and almost a similar mean radius over time. The Sc content in 6XXX_3 is almost double that of 6XXX_1, so we checked the variation of volume fraction over time for these alloys.

The variation of volume fraction for the three alloys has been shown in Figure 8.5. Here, one can observe that the volume fraction increases with an increase in Sc content.

In Figure 8.4, the mean radius of the Al₃Sc phase for alloys 6XXX_1 and 6XXX_3 is almost identical, but the volume fraction of the Al₃Sc phase for 6XXX_3 is significantly larger than that of 6XXX_1. Thus, we investigated the number density of these alloys and the results are presented in Figure 8.6.

In Figure 8.6, we can observe that number density curves seem to overlap for 6XXX_1, 6XXX_2, and 6XXX_3. Thus, we have presented a magnified view of the number density curves. From the magnified view, one can observe that the number density of the Al₃Sc phase of 6XXX_1 is lower than that for 6XXX_3. Thus, now it

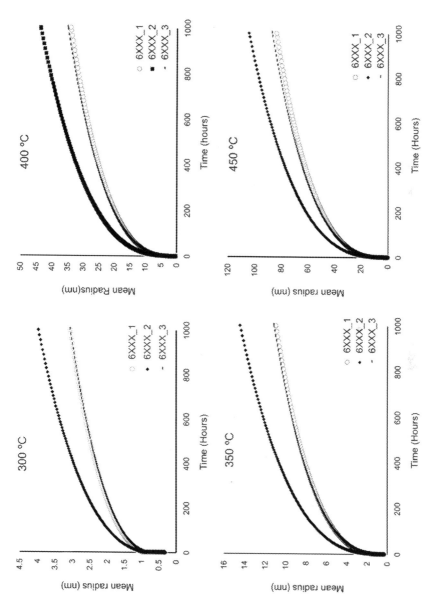

FIGURE 8.4 Mean radius (nm) vs time (hours) for isothermal heat treatment for 6XXX_1, 6XXX_2, and 6XXX_3.

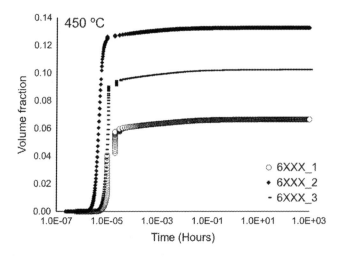

FIGURE 8.5 Volume fraction vs time (hours) for isothermal heat treatment for 6XXX_1, 6XXX_2, and 6XXX_3.

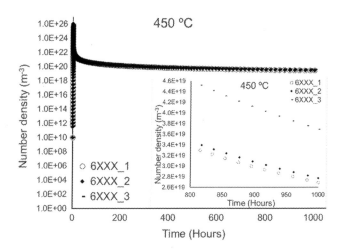

FIGURE 8.6 Number density vs time (hours) for isothermal heat treatment for 6XXX_1, 6XXX_2, and 6XXX_3.

is understandable that even though the mean radius Al_3Sc phase is almost identical for 6XXX_1 and 6XXX_3, the volume fraction of the Al_3Sc phase is significantly larger for 6XXX_3 when compared with 6XXX_1. Thus, an increase in Sc leads to an increase in the number density of Al_3Sc grains in these alloys.

8.4   CONCLUSIONS

The novelty of this work can be summarized in a few points as follows:

- Documentation on aluminum alloys using the CALPHAD approach generally concentrates on compositions that are around the familiar compositions of standard alloys (Assadiki et al. 2018). As part of this work, we developed a framework to predict and test new compositions through in-depth research based on CALPHAD and artificial intelligence.
- Aluminum alloys with 6–8 alloy elements have been studied through the CALPHAD approach (Andersen et al. 2018, Assadiki et al. 2018, Haidemenopoulos et al. 2010). We demonstrated that alloys with 10–12 alloying elements can be efficiently investigated in the Thermocalc software. This will allow researchers to apply this approach in their own work, allowing them to add new elements to their existing alloys.
- The Scheil-Gulliver solidification simulation provides vital information on precipitation of the Al_3Sc phase. It shows the effect of the amount of Sc on liquid temperature.
- The precipitation kinetics simulation provides vital information on grain growth and coarsening as a function of Sc content in the alloy. Additionally, it is important to examine volume fraction and number density while performing a heat-treatment simulation.

In the future, plans are to perform heat-treatment simulations on other nucleation sites like grain boundaries and dislocations. Heat-treatment simulations require optimized values of interfacial energy between the precipitate and the matrix phase in order to mimic experimental findings, which is a must from an application point of view. Thus, our planned future work will focus on optimizing interfacial energy at various nucleation sites for multi-component aluminum alloys with a scandium addition.

9 Case Study #9: Titanium Alloys for High-Temperature Applications: Combined Machine Learning-CALPHAD Approach

9.1 PROBLEM FORMULATION

In this chapter, we use data for the Ti-Al-Cr-V system, where aluminum varies from 0 to 50 weight % and chromium and vanadium vary between 0 and 15 weight %. Aluminum is a α-phase stabilizer, while Cr and V are β-phase stabilizers. Commercially available titanium alloys can be predominantly α-phase or β-phase, or a mixture of both α-phase or β-phase. The Al addition above 9–10% makes Ti-based alloys predominantly α-titanium alloys. Due to the two β-stabilizers, Cr and V, we are dealing with a (α + β) titanium alloy system. Additionally, we have Young's modulus and density values calculated at temperatures between 30 and 1500°C by using a commercial software package called JMatPro and optimization package IOSO. We perform equilibrium calculations for these 102 alloys at temperatures (30–1500°C) through the Thermocalc software using the TCTI2 Thermodynamic database. We obtain values for stable phases, fraction of phases, and a fraction of alloying elements in these phases for all the candidate alloys. Even though we are dealing with a (α + β) titanium alloy system, there exist several cases where titanium aluminide phases are stable. For a (α + β) titanium alloy system, titanium aluminides are detrimental, but recently, a significant amount of research has been performed on titanium aluminides within the framework of the CALPHAD approach.

All of the data are analyzed through the concept of self-organizing maps (SOMs) for understanding various patterns and meaningful correlations. Finally, we list a set of alloy candidates for titanium aluminides that are expected to be stable at temperatures above 1000°C. We list the composition, temperature, stable phases, and Young's modulus of elasticity and density of these alloys.

DOI: 10.1201/9781003167372-9

9.2   CALPHAD APPROACH: THERMODYNAMICS/PHASE TRANSFORMATION

In this section, we state the phase notations from Thermocalc. Titanium alloys consist of several phases: HCP_A3 (α) and BCC_B2 (β), titanium aluminides, like ALTI3 _D019 (α_2), ALTI_L10 (γ), AL2TI, AL5TI2, AL5TI3, and C15_LAVES. In Thermocalc, BCC_B2 (β) exists in two forms: as an ordered and a disordered phase, which can be displayed on the phase diagram. The B2 phase exists in two forms: BCC_B2 and BCC_B2#2. Titanium aluminides (TI3AL _D019 (α_2) and TIAL_L10 (γ)) are known for high-temperature strength and comparatively lower density. It is used in the aerospace industry as entirely Ti_3Al ($\alpha2$) and γ-TiAl ordered L10 structure (ALTI_L10) compounds. The ALTI3_D019 (α_2) phase is experimentally observed for Al concentration above 9%, but these precipitates are brittle and reduce the fracture strength and ductility. In this work, we analyze the ($\alpha + \beta$) titanium alloy system. Additionally, we explore compositions that will provide titanium aluminide phases (TIAL_L10 (γ)).

9.2.1   Final Data from CALPHAD That Will Be Used for SOM Analysis

The data table has 102 rows and 57 columns.

- 102 rows are for 102 alloys.
- 57 columns: 4 alloying elements, 1 temperature, 10 equilibrium phases, 40 alloying element fractions in an equilibrium phase, and 2 properties (Young's modulus and density of the alloy).

This table has numerous blank cells as not all equilibrium phases are stable for each candidate alloy. These data are analyzed using the self-organizing maps (SOMs) algorithm, which is known to capture the topology of multivariate data sets.

9.3   APPLICATION OF MACHINE LEARNING ALGORITHMS

The SOM algorithm has been discussed in detail in Chapter 3. The distribution of aluminum in the SOM space is shown in Figure 9.1. One can observe that there are squares inscribed in the hexagonal units of varying size, while for some units there is a dot. The size of the square is a measure of the number of candidates that are positioned on the edges of the hexagonal units. The larger the number of candidates, the larger the square. A dot means that no candidates are positioned on the edges of that particular hexagonal unit. This can be better understood through Figure 9.2, where the distribution of chromium is shown. On Figure 9.2, all of the candidate alloys are shown to illustrate the meaning of the squares inscribed within the hexagonal units.

Adding all the candidates on the figure can be done for small data sets. For larger data sets, it will be difficult to understand these figures in the presence of all the numbers written on the figure. Thus, in all the other figures, we have identified only

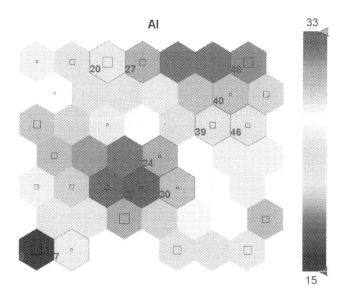

FIGURE 9.1 SOM analysis: Distribution of aluminum (wt. %).

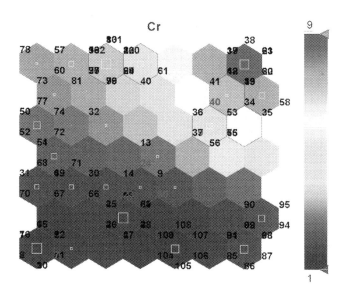

FIGURE 9.2 SOM analysis: Distribution of chromium (wt. %).

a few units that possess the best-performing candidates. These diagrams need to be read along with other figures for a better understanding of the alloy system.

Figure 9.3 shows the distribution of Young's modulus on the SOM space. For high-temperature applications it is desired that the alloy must possess a high value for Young's modulus. Thus, unit numbers 20, 27, 48, and 0 need to be analyzed.

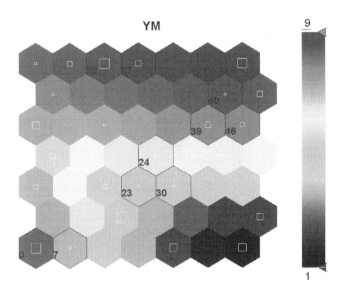

FIGURE 9.3   SOM analysis: Distribution of Young's modulus.

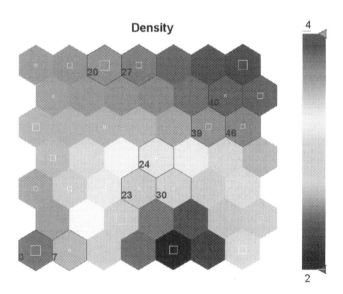

FIGURE 9.4   SOM analysis: Distribution of density.

Figure 9.4 shows the distribution of density in the SOM space. From an application point of view, a lower density is desired. Thus, unit numbers 23, 24, and 30 need to be analyzed, but the Young's modulus associated with these units is quite low. One needs to be extremely cautious with density, as we have analyzed candidate alloys for high-temperature applications (up to 1500°C). There are several

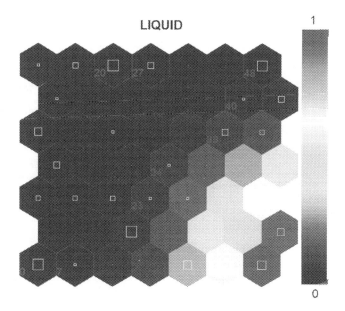

FIGURE 9.5 SOM analysis: Distribution of LIQUID phase (mole %).

cases where the LIQUID phase appeared as shown in Figure 9.5. Since it is a LIQUID phase, both Young's modulus and density will be low for alloys included in these units. Thus, for an optimum combination of Young's modulus and density, one needs to find a compromise between the two and work in the units in the vicinity of unit numbers 20, 27, 48, and 0.

Figure 9.6 is extremely useful if one wants to work with an α-titanium alloy or β-titanium ally or a combination of both. One can observe that unit 0 corresponds to a low temperature, for which the stable phase is HCP_A3. HCP_A3 is a low-temperature phase, and SOMs have been able to position alloys containing a high amount of α-phase (HCP_A3) in the low-temperature region. If one wants to design an alloy for a high-temperature application, they can work with the BCC_B2 phase (units 20 and 27). Similarly, if one wants to design an alloy for medium-temperature application, they can work with the BCC_B2_2 phase (units 40 and 48) (Figure 9.6).

9.4 TITANIUM ALUMINIDES

9.4.1 Conventional Experimental Design

There are several approaches that are employed by experimentalists in designing the chemistry of titanium aluminides. Two of them are listed below:

- Titanium aluminides: γ-TiAl alloys as well as (γ-TiAl + α_2-Ti_3Al) alloys can be synthesized in a composition range Ti– (46–52) Al– (1–10) M (at. %), where M is one of the following transition elements: V, Cr, Mn, Nb, Mn, Mo,

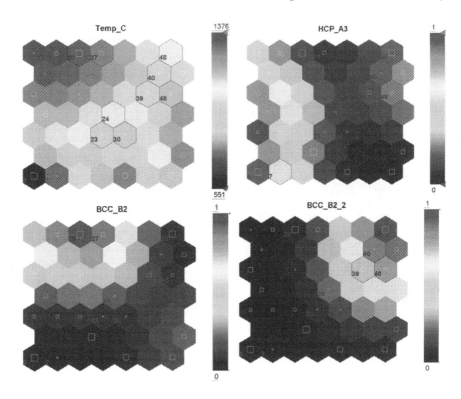

FIGURE 9.6 SOM analysis: Distribution of temperature (C), and BCC_B2, HCP_A3, and BCC_B2_2 phase (mole %).

and W. Reported density values of these alloys are in a range, 3.9–4.1 gcm^{-3}, whereas the density of γ-TiAl reported is about 3.4 gcm^{-3}.

- In another approach, the element added for improving mechanical properties is divided into a few categories: X_1, X_2, and X_3, where alloy composition can be written as Ti-(45–49) Al-(1–3) X_1-(1–4) X_2-(0.1–1). X_3 is added in the case of (γ-TiAl + α_2-Ti$_3$Al) alloys:
  1. X_1: Mn, Cr, Mo
  2. X_2: Nb
  3. X_3: C, N, Si, W, B, Re

9.4.1.1 Challenges

By using the two approaches to design chemistry, a large number of samples have to be manufactured and tested to demonstrate improvements. For a given chemistry, the desired properties can be significantly improved with the development of manufacturing protocols. Manufacturing protocols include additive manufacturing as well as thermal treatment of samples to stabilize γ-TiAl in a targeted grain size and volume fraction. After fabrication, samples are tested and characterized to verify improvements. This route will be both time consuming and expensive.

9.4.2 Proposed Method

Titanium aluminides are for applications at temperatures in the vicinity of 1000°C. Choosing the correct element for addition as well as the chemistry optimization will be a complex task. A combination of CALPHAD and machine learning is proposed. Phase stability of γ-TiAl was estimated under the framework of CALPHAD.

Figures 9.7 and 9.8 show the distribution of the ALTI_L10 and ALTI3_D019 phases in the SOM space. In Figure 9.7, the highest amount of ALTI_L10 is around unit 23. A LIQUID phase exists around this unit, and it is expected as we want to analyze candidate alloys up to 1500°C. At 1500°C, some alloys start melting while some stay solid.

In Figure 9.8, unit 0 contains an alloy with the highest ALTI3_D019. Now, unit 0 is a low-temperature region and also contains an alloy with the highest amount of HCP_A3 phase. During data analysis, we have observed only one candidate that has a fully stabilized ALTI3_D019 phase at 1000°C.

Hence, we will focus on ALTI_L10 for determining composition that will provide a fully stabilized ALTI_L10 phase. Figure 9.9 is important for studying the ALTI_L10 phase. If we just analyze ALTI_L10 and temperature, we notice the area around unit 23 is a region of average temperature. But, during the phase stability calculation, several candidates are observed as fully ALTI_L0 stabilized (mole % is 1). A few candidate alloys are listed in Table 9.1. Therefore, we analyze the phase content and find that the chromium content in ALTI_L10 is around the mean values in cells 40 and 48. Thus, there must be some ALTI_L10 in candidate alloys in those units for it to show up in the SOM analysis.

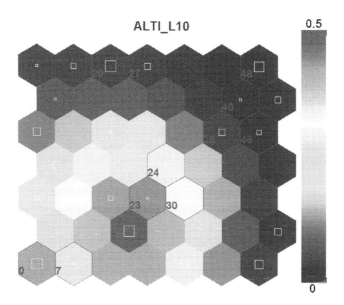

FIGURE 9.7 SOM analysis: Distribution of ALTI_L10 phase (mole %).

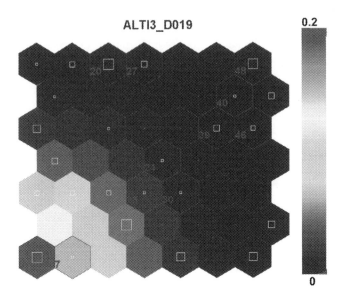

FIGURE 9.8   SOM analysis: Distribution of ALTI3_D019 phase (mole %).

The SOM plot must be analyzed in groups for understanding patterns in a data set. In this case, we have 57 components. One hundred and two candidates are positioned at the vertices of each unit. The average value over a unit for any property is denoted by color. Thus, one must compare a SOM plot with a number of other plots. If a user thinks that the correlations are not as per their understanding of the system, they must change the parameters and train the model again.

Table 9.1 shows the amount of the γ-TiAl (ALTI_L10) phase for a set of compositions and temperatures above 1000°C. In Table 9.1, we have used the notations as it exists in the Thermocalc database so that readers can correlate it with the manual and literature. Apart from titanium and aluminum, we use chromium and vanadium as elements that need to be added. We can observe from Table 9.1 that these candidates have densities in the range of 3.0892-3.8984 * 10^3 Kg/m^3. Chromium and vanadium improve ductility but reduce oxidation resistance. Since the calculated density is low, we can use Nb or Ta as additional elements.

The purpose of Table 9.1 is to demonstrate the stability of γ-TiAl at elevated temperatures for various compositions. It can be observed that several other phases exist in a few cases. Thus, in order to design an alloy containing the γ-TiAl phase, one needs to be careful in choosing the composition, as other phases can prove to be detrimental when this alloy is used for applications at elevated temperatures.

Through the CALPHAD approach, one can also scan through various elements reported in literature as well as other elements in the database that have not been used before and study their effect on the stability of the γ-TiAl phase. Manufacturing and heat-treatment protocol play important roles in achieving a desired microstructure that will be helpful in achieving the desired properties.

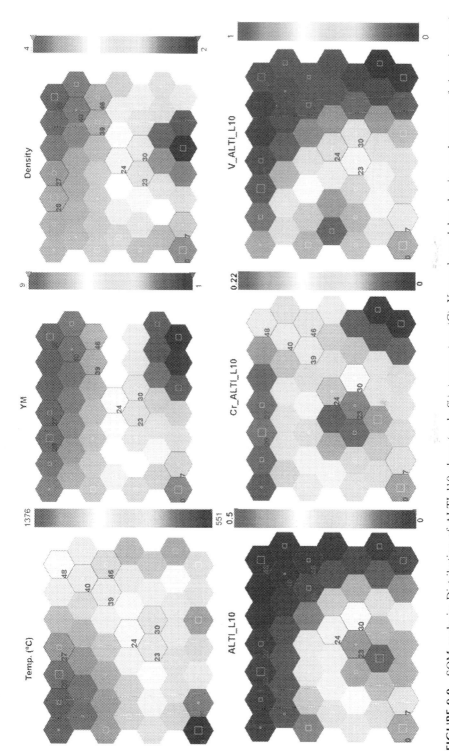

FIGURE 9.9 SOM analysis: Distribution of ALTI_l10 phase (mole %), temperature (C), Youngs's modulus, density, and amount of chromium and vanadium in ALTI_L10 (g/mol).

TABLE 9.1
Stability of γ-TiAl (ALTI_L10) Phase

#	Temp. (°C)	Composition (wt. %)				JMATPro		Thermocalc			
		Ti	Al	Cr	V	Density*10³ (kg/m³)	Y.M.*10⁸ (N/m²)	BCC_B2	HCP_A3	AL5TI2	ALTI_L10
1	1200	45.385	50	4.5147	0.1	3.1035	1.5641			0.2818	0.71692
2	1200	46.751	49.979	3.1699	0.1	3.1017	2.6169			0.3726	0.62743
3	1200	49.8	50	0.1	0.1	3.0892	4.9269			0.6318	0.36824
4	1200	45.501	47.569	0.1466	6.7834	3.2796	5.3638				1
5	1200	58.036	41.764	0.1	0.1	3.2427	5.1815				1
6	1200	61.826	37.959	0.1149	0.1	3.3717	5.7998				1
7	1200	66.329	33.144	0.1	0.4275	3.5755	6.9432		0.12222		0.87778
8	1200	42.651	33.674	10.764	12.911	3.7853	7.5887	0.50469			0.49531
9	1200	46.674	30.252	10.234	12.84	3.8984	9.6086		0.29486		0.14486
10	1300	52.435	46.309	1.1563	0.1	3.1285	2.9695				1
11	1300	52.435	46.309	1.156	0.1	3.1285	2.9704				1
12	1300	48.559	45.526	1.0426	4.8726	3.262	1.9691				1
13	1300	54.506	39.215	0.1139	6.165	3.4406	5.4923				1
14	1300	60.179	38.048	1.6728	0.1	3.4485	6.0925				1
15	1300	66.9	31.91	1.0902	0.1	3.5825	6.3053		0.88465		0.11535
16	1400	57.745	41.855	0.2995	0.1	3.2524	0.5655				1
17	1400	59.533	39.518	0.5412	0.4077	3.3578	3.2351				1
18	1400	61.011	37.708	0.7461	0.5356	3.4231	5.7457		0.20081		0.79919
19	1400	62.85	35.351	0.9683	0.8307	3.479	5.7058		0.80583		0.19417
20	1400	43.104	41.632	0.2768	14.988	3.3345	0.4249		0.27813		0.72187
21	1400	44.47	40.281	0.2589	14.99	3.4016	1.4418		0.52518		0.47482
22	1400	46.708	38.121	0.1757	14.995	3.499	5.9747		0.99107		0.00893

9.5 CONCLUSIONS

- The CALPHAD and AI approach can be helpful in determining complex correlations in highly complex phase stability data.
- In this work, we list the ways by which a reader can design a wide range of alloys for Ti-based alloys, such as α-titanium, β-titanium, or a combination of both. Additionally, a user can attempt to design titanium aluminides.

Thus, the most important part is that using the same data set, a reader can design a completely different alloy. In this work, we analyzed the data after one analysis, and changed the problem formulation for designing another class of titanium alloys from the same data set. This data set was initially generated for a (α + β)-titanium alloy. By setting a new goal, we have tabulated a set of compositions and temperatures that will provide a fully stabilized ALTI_L10 phase. We have drawn meaningful correlations that will be helpful in designing manufacturing protocols of these alloys.

10 Case Study #10: Design of β-Stabilized, ω-Free, Titanium-Based Biomaterials: Combined Machine Learning-CALPHAD Approach

10.1 CALPHAD APPROACH: THERMODYNAMICS/PHASE TRANSFORMATION

The main focus of this chapter is the discovery of a new composition for Ti-based biomaterials. As stated in Chapter 2, Ti-based biomaterials must have a fully stabilized β-phase. These alloys must be free from any other detrimental phase, particularly the ω-phase. We analyzed a Ti-Ta-Nb-Sn-Mo-Zr system for studying stability of α-, β-, α2-, and ω-phases. Twenty-six sub-groups were formed: three elements together, four elements together, five elements together, and the final group with six elements. All 26 sub-groups were studied for determining phase stability of the omega (ω)-phase from 30 K (−242°C) to 1504 K (1231°C), and the ω-phase was observed up to 1149 K (876°C). Thus, we accumulated a large amount of data for study.

In a previous chapter on Ti-based alloys for high-temperature applications, we started with working on a system that will be a combination of α-Ti alloy and β-Ti-alloy. But, we were able to use the same data and determine new composition for stabilizing titanium aluminides.

In this work, too, our main focus was Ti-based biomaterial. We are working on minimizing a complex problem of determining novel compositions that will be helpful in developing a ω-phase free titanium-based biomaterial. And we arranged lots of data as per our problem formulation: maximize the β-phase and minimize the α-, α2-, and ω-phases.

So, we looked at what other problems we can solve with this data. From literature and Chapter 2, we know that the ω-phase is problematic for another important Ti-based system, shape memory alloys (SMAs). There exist several classes of SMAs that can be studied with the data generated for the Ti-Ta-Nb-Sn-Mo-Zr system. Ti-based biomaterials used for implants must not contain an α2-phase, but an α2-phase is important for shape memory applications. SMAs can contain

DOI: 10.1201/9781003167372-10

TABLE 10.1

Upper and Lower Bounds of Design Variables

	Temperature		Mo	Nb	Ta	Zr	Sn	Cr	V	Fe	Ti
Unit	K	°C					Mole %				
Min.	30	−242	0	0	0	0	0	0	0	0	Balance
Max.	1504	1231	9	32	24	10	5	5	5	5	

elements like Fe, Cr, and V, so we expanded our research and created another database. This revised system contains Ti-Ta-Nb-Sn-Mo-Zr-Fe-Cr-V in the proportions given in Table 10.1.

10.2 APPLICATION OF MACHINE LEARNING ALGORITHMS FOR PREDICTING NOVEL COMPOSITIONS/TEMPERATURE FOR IMPROVING PHASE STABILITY OF DESIRED β-PHASE WHILE AVOIDING DETRIMENTAL α- AND ω-PHASES

10.2.1 Problem Formulation

The alloy design was performed in two stages.

10.2.1.1 Ti-Ta-Nb-Sn-Mo-Zr-Fe-Cr-V System

In this work, we have analyzed the-Ta-Nb-Sn-Mo-Zr-Fe-Cr-V system and determined the amount of the ω-phase between 30 K (−242°C) to 1504 K (1231°C). This system contains ten variables (nine alloying elements and temperature) and one objective (ω-phase). This data was analyzed through several concepts of artificial intelligence. The purpose was to determine the alloy system that is expected to contain the least amount of ω-phase.

Various approaches utilized can be summarized as follows:

- Statistics or correlation analysis: Figure 10.1 shows the correlation plot for the design variables and objectives. Regarding OMEGA (ω-phase), it is inversely affected by temperature. We know from literature that OMEGA (ω-phase) is stable at cryogenic temperatures. The amount of OMEGA (ω-phase) decreases with a rise in temperature, and it is not stable at room temperature. But, OMEGA (ω-phase) is present in microstructures and is detrimental for aerospace and biomedical applications. Other correlations are weak. Titanium enhances formation of OMEGA (ω-phase). Niobium (Nb) and tantalum (Ta) inversely affect OMEGA (ω-phase), but correlations are too weak. The correlation of OMEGA (ω-phase) with other elements is even weaker. Weak correlations were expected as there are ten design variables. There are 3,000 rows and 11 columns; one of those columns is OMEGA (ω-phase). Calculations were performed between 30 K

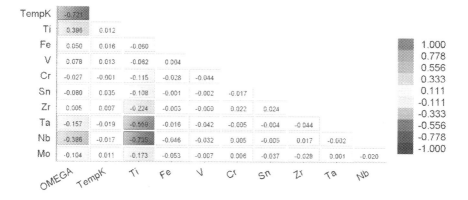

FIGURE 10.1 Correlation matrix.

($-242°C$) to 1504 K ($1231°C$), and the ω-phase was observed up to 1149 K ($876°C$). Thus, there are lots of missing values for a number of rows. Thus, weak correlations are expected.

- Supervised machine learning (ML): We used a software, RapidMiner, for developing predictive models for the ω-phase. The best-performing algorithms are decision tree and gradient boosted trees. The table provides information on errors associated with all the other algorithms. A list of algorithms used for this analysis is as follows:
  - Generalized Linear Model
  - Deep Learning
  - Decision Tree: Best performer
  - Random Forest
  - Gradient Boosted Trees: Best performer
  - Support Vector Machine
- Unsupervised machine learning: Two different approaches were used:
  - Self-Organizing Maps (SOMs): It was used for determining correlations between OMEGA (ω-phase) and the other ten design variables.
  - Principal Component Analysis: PCA analysis was used for determining a few elements that affect OMEGA (ω-phase) inversely. Most promising elements were identified as niobium (Nb) and tantalum (Ta) for this database.
- Multi-objective optimization: After identifying the Ti-Nb-Ta system, we went ahead with performing multi-objective optimization through evolutionary algorithms (EAs) or genetic algorithms (GAs). Problem formulation: maximize β-phase and minimize α-, $α_2$-, and ω-phases.

10.2.1.2 Ti-Nb-Ta System

Titanium-based biomaterials of this system (Ti-Nb-Ta) already exist. In a recent work (Ferrari et al. 2019), the authors performed high-throughput DFT simulations to demonstrate the possibility of finding OMEGA (ω-phase)-free alloys for Ti-Ta-X system, where X is Sn and Zr. From CALPHAD analysis, we have data for all of these systems.

10.3   SUPERVISED MACHINE LEARNING IN RAPIDMINER

We worked on the-Ta-Nb-Sn-Mo-Zr-Fe-Cr-V system in this case study and developed models for the OMEGA phase. Error metrics of these models are listed in Table 10.2. From Table 10.2, one can observe that "decision tree" and "gradient boosted trees" were the best performers. The true values of the OMEGA phase were plotted with the predicted values. Figure 10.2 shows the plot for the decision tree algorithm, while Figure 10.3 shows the plot for the gradient boosted trees algorithm (Figure 10.4).

TABLE 10.2
Error Metrics for Supervised Machine Learning Models Developed in RapidMiner

Model	Root Mean Squared Error	Standard Deviation	Total Time (s)
Generalized Linear Model	0.1210	0.0022	1.9
Deep Learning	0.0733	0.0078	4.5
Decision Tree	0.0860	0.0093	1.1
Random Forest	0.0825	0.0027	10.1
Gradient Boosted Trees	0.0595	0.0058	17.8
Support Vector Machine	0.0908	0.0109	81.8

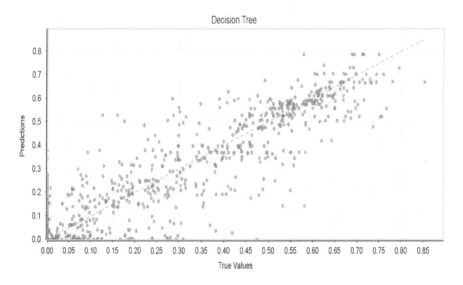

FIGURE 10.2   Decision tree: True values vs. predicted values for OMEGA phase.

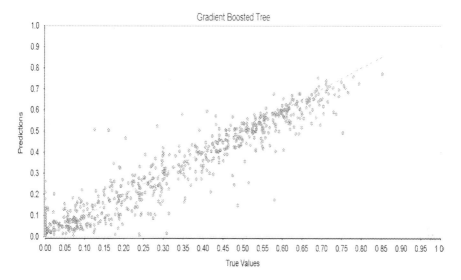

FIGURE 10.3   Gradient boosted trees: True values vs. predicted values for OMEGA phase.

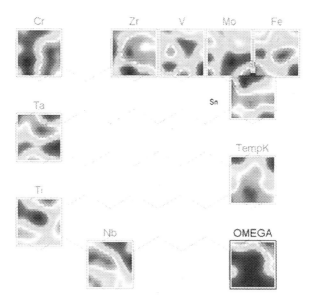

FIGURE 10.4   SOM component plot for the Ti-Ta-Nb-Sn-Mo-Zr-Fe-Cr-V system.

Predictive models developed through gradient boosted trees performed best in this analysis. From Table 10.2, one can observe that these models were trained in 17.8 seconds. Training time is quite fast for all the models as we are dealing with a complex data set.

In RapidMiner, single objective optimization can be performed. In our problem formulation, we have four objectives. Therefore, models developed in RapidMiner can be used as a predictive tool. In cases where there is just one objective, within a few minutes a user can perform all the tasks just by a few clicks. No programming is needed if a user does not want to code. There are plenty of tools for visualization in RapidMiner.

10.4   UNSUPERVISED MACHINE LEARNING IN modeFRONTIER: SELF-ORGANIZING MAPs (SOMs)

Figure 10.4 shows the SOM component plot for Ta-Nb-Sn-Mo-Zr-Fe-Cr-V system. SOM algorithm settings are as follows: X-Dimension = 15, Y-Dimension = 18, and total map units = 270. The neighbor's function is Gaussian. A few other model parameters were varied for minimizing errors associated with the models. The quantization error is 0.100 and the topological error is 0.076. Even though the quantization error is high at 10%, the topological error is about 7.6%. It is recommended that this model must not be used for prediction. But, the SOM algorithm is quite efficient in preserving the topology of the data set. In this case, an error of 7.6% is acceptable as the data set has 3,000 rows and 11 columns, with lots of missing points (Figure 10.4).

Figure 10.5 shows the distribution of a few features associated with the Ta-Nb-Sn-Mo-Zr-Fe-Cr-V system on the SOM space. On close analysis of all the plots, it was observed that most of the correlations were mixed. Only Nb and Ta show some direct relation with temperature and OMEGA phase. That is in a region where the average temperature is around 400 K and above, the highest amount of OMEGA phase is observed. And in these regions, Nb and Ta is the lowest. This correlation will be kept in mind in the next part; that is, principal component analysis (PCA).

10.5   UNSUPERVISED MACHINE LEARNING IN modeFRONTIER: PRINCIPAL COMPONENT ANALYSIS (PCA)

Figure 10.6 shows the 2-D scatter plot for the Ti-Ta-Nb-Sn-Mo-Zr-Fe-Cr-V system. PC1 and PC2 are the first two principal components obtained during PCA analysis. PCA analysis can be helpful in determining the relation between OMEGA phase, temperature, and nine alloying elements. PCA scores or projections are plotted in the reduced 2-D space. On the figure, few arrows are visible, while some arrows are extremely small. Along any principal component (PC1 or PC2), the length of the arrow is a measure of the relative contribution of that variable along that PC.

In Figure 10.6, there are five long arrows: OMEGA, temp (K), Ti, Ta, and Nb. This analysis was performed for determining correlations between OMEGA phase, and alloying elements in particular. Temperature and titanium cannot be removed from this analysis as this alloy is titanium based. Thus, we have to choose among the other eight elements. One can see that Ta and Nb arrows are longer when compared with other element arrows. Arrows for both Nb and Ta are in the opposite direction of OMEGA; thus, PCA analysis shows that an increase in Nb and Ta will lead to a decrease in OMEGA phase.

A 3-D scatter plot was plotted in Figure 10.7 and its enlarged view can be seen in Figure 10.8. The first three components were used: PC1, PC2, and PC3. In these

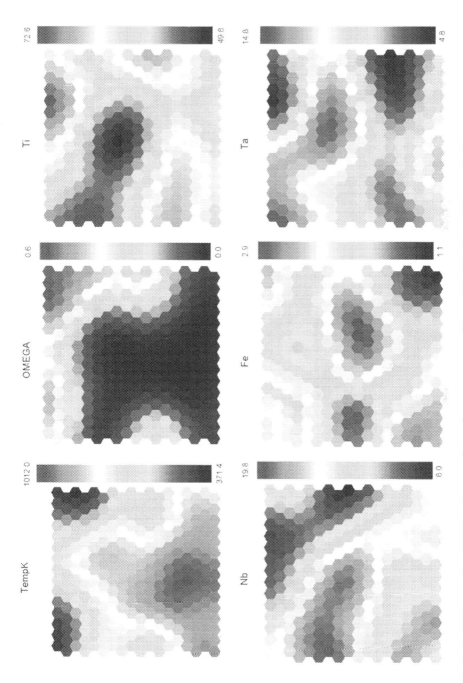

FIGURE 10.5 SOM plots for OMEGA phase (mole fraction); temperature (K); and Nb, Fe, Ti, and Ta in mole %.

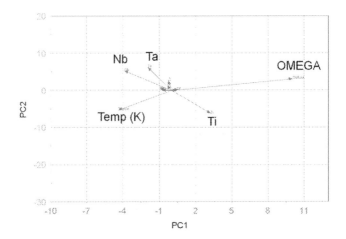

FIGURE 10.6   PCA analysis: 2-D scatter plot for the Ti-Ta-Nb-Sn-Mo-Zr-Fe-Cr-V system.

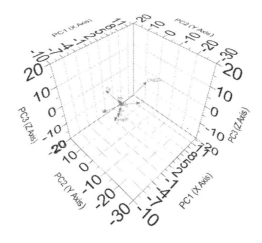

FIGURE 10.7   PCA analysis: 3D-scatter plot.

figures, too, one can observe that among the eight alloying elements, Nb and Ta have the longest arrow, thus the highest contribution.

Thus, we selected the Ti-Nb-Ta system for further analysis.

10.6   MULTI-OBJECTIVE OPTIMIZATION IN modeFRONTIER

10.6.1   VARIABLE SCREENING

Variable screening was performed prior to the development of the predictive models in modeFRONTIER software. The relative contributions of the input variables are derived using the smoothing algorithm of the ANOVA curve (Table 10.3).

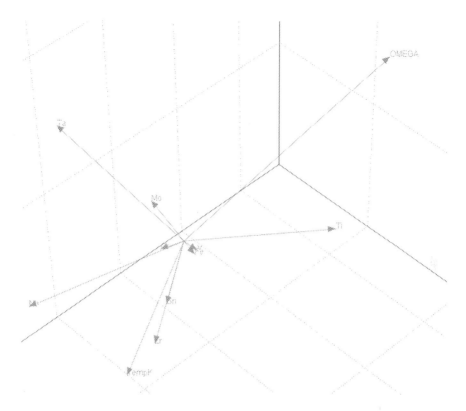

FIGURE 10.8 PCA analysis: Enlarged view of 3-D scatter plot.

TABLE 10.3
Relative Contribution of the Design Variables along with Performance Index

	BCC_B2	HCP_A3	OMEGA	ALTI3_D019
Ti	0.022	-0.023	0.093	0.021
Nb	0.047	0.079	0.050	0.047
Ta	0.013	0.024	0.061	0.010
Temp (K)	0.918	0.920	0.796	0.923
Mean Absolute Error	0.051	0.055	0.085	0.057
Mean Normalized Error	0.051	0.055	0.085	0.058
R-squared	0.947	0.939	0.867	0.939

10.6.2 DEVELOPMENT OF SUPERVISED MACHINE LEARNING MODELS

Predictive models were developed for BCC_B2, HCP_A3, OMEGA, and ALTI3_D019 as a function of Ti, Nb, Ta, and temp (K). The K-NN (K-nearest neighbor) algorithm was chosen for developing the predictive models. Model performances are listed in Table 10.4.

TABLE 10.4

Error Metrics of Predictive Models Developed in modeFRONTIER

	BCC_B2	HCP_A3	OMEGA	ALTI3_D019
Mean Absolute Error	0.026	0.0299	0.0278	0.0285
Mean Normalized Error	0.027	0.0311	0.0302	0.0302
R-squared	0.97	0.96	0.963	0.957
AIC	−1680	-1590	−1710	−1580

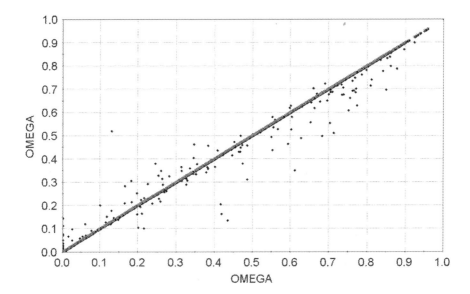

FIGURE 10.9   OMEGA phase in Ti-Nb-Ta system: True values vs. values predicted by K-NN models.

Figure 10.9 shows the comparison plot for the true values of OMEGA phase versus the values predicted through K-NN models.

One can refer to Figures 10.10 and 10.11 for understanding the distribution of OMEGA phase at 300 K and 800 K. The titanium content was fixed at 71 mole % for this case. Fine black dots are visible in Figures 10.10 and 10.11. These represent the candidate alloys.

In Figure 10.12, titanium content was fixed at 55 mole % and temperature is 300 K. There exists a large region on the diagram where OMEGA phase is stable.

From Figures 10.10 and 10.12, one can observe that a significant amount of OMEGA phase is present at 300 K. In thermodynamic calculations, one will not observe any trace of OMEGA phase at 800 K. In Figure 10.11, one can observe that there exists a region where OMEGA phase is negligible for Ta + Nb = 29 mole %. But, for this composition, one will not be able to manufacture an OMEGA-free alloy at 300 K.

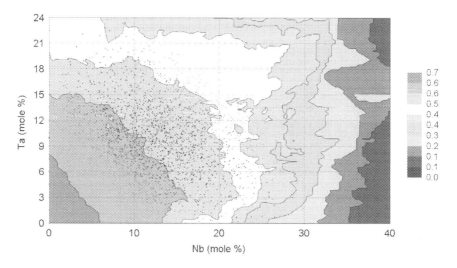

FIGURE 10.10   Distribution of OMEGA phase at temperature (300 K) and Ti (71 mole %).

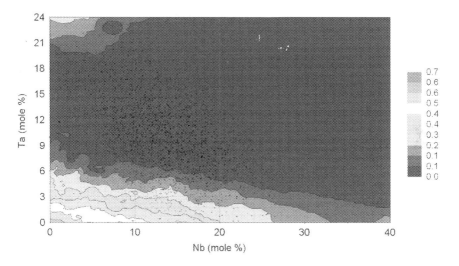

FIGURE 10.11   Distribution of OMEGA phase at temperature (800 K) and Ti (71 mole %).

Even in Figure 10.12, for Ta + Nb = 45, one cannot expect an OMEGA-free alloy at 300 K.

This information will be helpful while performing multi-objective design optimization.

10.6.3   MULTI-OBJECTIVE OPTIMIZATION

Figure 10.13 shows the workflow created in modeFRONTIER for optimization. In the figure, it is shown that FAST-MOPSO is the optimizer. We have added a description of all these algorithms in Chapter 2.

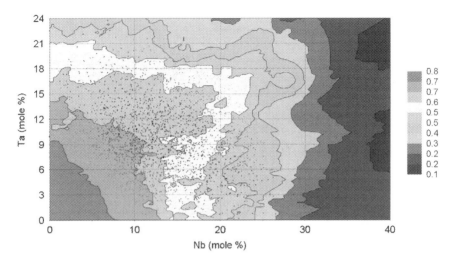

FIGURE 10.12　Distribution of OMEGA phase at temperature (300 K) and Ti (55 mole %).

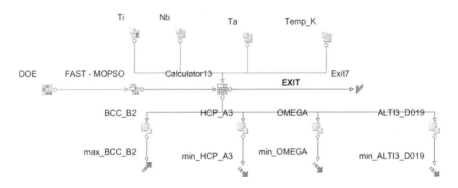

FIGURE 10.13　Workflow created in modeFRONTIER for optimization.

Problem formulation: maximize β-phase (BCC_B2) and minimize α (HCP_A3), α_2(ALTI3_D019), and ω-phase (OMEGA) (Figure 10.13).

During optimization, a large amount of data was generated. It will be shared along with this book.

10.7　CONCLUSIONS

In this case study, we have focused on discovering an OMEGA-free system of Ti-based biomaterials.

Key points can be listed as follows:

- Selected a set of non-toxic elements to develop a database for the Ti-Ta-Nb-Sn-Mo-Zr system.

- Expanded this database by adding Fe, Cr, and V and determined phase stability of OMEGA phase.
- Through supervised and unsupervised ML, determined that Ti-Ta-Nb is the most appropriate system with respect to minimizing the amount of OMEGA phase.
- Utilized this data from the library created in the beginning.
- Developed predictive models for various phases as a function of composition and temperature.
- Formulated multi-objective optimization problem for developing OMEGA-free Ti-Nb-Ta alloys.

Elements like Fe, Cr, and V were added so that a user can utilize this database and this methodology for developing OMEGA-free shape memory alloys, although several shape memory alloys can be studied by analyzing the Ti-Ta-Nb-Sn-Mo-Zr system.

11 Case Study #11: Industrial Furnaces I: Application of Machine Learning on Industrial Iron-Making Blast Furnace Data

11.1 BLAST FURNACE IRON-MAKING

In Chapter 2, we have described the blast furnace iron-making process in detail. In this chapter, we have described the blast furnace operation with respect to the case studies. This chapter is a combination of three (3) different case studies. In this chapter, data from three different blast furnaces were acquired from three different integrated steel plants. These steel plants are at different geological locations.

From Chapters 4 to 9, these chapters include case studies that can be performed in a closed environment. Topics include alloy design, nanoindentation experiment, imaging through an atomic force microscope and analysis in atom probe tomography. A researcher can be present at any geographic location on this planet, but still can acquire resources needed for these case studies from the same vendor and develop similar laboratory and expect similar results.

In the case of blast furnaces in this chapter or LD furnace in Chapter 12, an operator cannot expect similar results that were achieved at an integrated steel plant in a different geographical location. The weather pattern is different. They use different raw materials. Additionally, one can observe in the later text that the furnace operating parameters were different, and the objectives were also different.

11.1.1 BLAST FURNACE OPERATION: ANALYTICAL MODEL AND DATA-DRIVEN MODEL

The blast furnace iron-making process is highly complex and non-linear in nature. Analytical models based on heat and mass balance are helpful to understand the process to some extent. Analytical models have been used for quantifying various parameters in the domain of metallurgical engineering (Dwivedi and Jha 2002). In Chapter 12, we have presented a case study on LD furnace where analytical models were developed through mass balance and works as per our expectation. Analytical models are based on the theoretical knowledge, like chemical reactions and heat and mass balance. These models are quite useful, but have limitations. This is mainly due to the fact that these models depend on many parameters that cannot be precisely determined. A blast furnace operates at about 1500°C. There are furnaces that can produce 10,000 tons of hot metal (molten iron) per day. The environment around the blast furnace and inside the blast furnace prevents the collection of the data needed for these models. At 1500°C, sometimes the sensors do not function properly and can give faulty reading. At times, there can be changes in the composition of raw materials, and input variables may affect each other too. All of this affects the accuracy of an analytical model of a blast furnace.

A data set obtained from an operating blast furnace is highly non-linear and noisy. A blast furnace can operate for a few years at a stretch without any shutdown. An analytical model or a statistical model may fit the data for 1 or 2 days, but in the long run, these conventional methods can be misleading, and these cannot be used as prediction models.

Supervised machine learning approaches like an artificial neural network are capable of fitting highly non-linear functions. It has been effectively used for fitting data from an operating blast furnace (Angstenberger 1996, Banks 1999, Jha et al. 2014, Mahanta et al. 2022).

Thus, in this work, we have applied artificial intelligence concepts for analyzing the three blast furnaces. We applied supervised and unsupervised machine learning algorithms for model development, analyzing patterns and determining correlations among process variables and objectives identified by that specific furnace operator.

11.1.2   CO$_2$ Emissions from a Blast Furnace

The steel sector accounts for approximately 5% of the world's energy consumption (Babich et al. 2002). Carbon dioxide (CO_2) is the main contributor among the gases responsible for the greenhouse effect. Greenhouse gas emissions from human activities are estimated at approximately 23 GB/A, with the steel industry contributing about 30%. About 60% of pig iron in the world is produced through blast furnaces (Schmöle and Lüngen 2004). The blast furnace, sintering plant, and coking plant consume about 70–75% of the total energy required for an integrated steelworks (Babich et al. 2002). Much of the fuel required to meet the above-mentioned energy requirements is covered by metallurgical coke (40–50%) (Babich et al. 2002). From an environmental perspective, it is estimated that approximately 4,150 kg of CO_2 is released to produce one ton of coke (Babich et al. 2002). The coke is then used as a fuel, thereby adding to the total CO_2 emissions of the integrated steelworks. Thus, CO_2 emissions are a major concern, and we have addressed it through one case study.

11.1.3   Hot Metal Temperature and Silicon Content in Hot Metal

Monitoring of hot metal temperature is extremely important in blast furnace operations. An operator can understand the thermal state of the hearth by monitoring hot metal temperature. A high temperature increases fluidity of the hot metal as it is helpful during tapping. Additionally, this hot metal is transferred to the downstream process so a high temperature will reduce the risk of solidification during transfer, but it will increase the fuel rate of the furnace. This extra energy requires burning of coke and coal, as it is the major fuel. Additional coke and coal translate to additional coke ash in the furnace. Coke ash is one of the major contributors of silicon content in hot metal.

A high silicon content in the hot metal will not only add to the production cost in steel making, it also gives an idea of the thermal state of the hearth. As silicon content in hot metal is closely associated with hot metal temperature, a high silicon content in hot metal means that the hearth is at a higher temperature than expected, thus it means that the coke (fuel) rate has gone up as it is responsible for heat requirements of the furnace. On the other hand, extremely low silicon content will correspond to lower hot metal temperature, which in other words means that the hearth is at a lower temperature than it must be. Unless the heat loss of the furnace is minimized, this lowering of hearth temperature may result in the chilling effect of the hearth, as the hearth must be maintained at a particular temperature range in order to ensure smooth operation of the furnace.

In order to ensure smooth operation of the furnace, it is recommended to keep the silicon within an optimal range.

11.2 DATASET USED IN THIS CHAPTER

We used data from three blast furnaces:

- Furnace 1: 2-month data
- Furnace 2: 1-month data
- Furnace 3: 1-year data

As mentioned, blast furnace operations vary from one plant to the other. So, the variables identified and recorded by the operator, and the problem that needs to be addressed, is decided by the operator and it differs from one plant to another. A data scientist must be comfortable working with data of various shapes and sizes. We have listed a number of techniques that will be helpful in analyzing blast furnace operations for data of different sizes.

11.3 FURNACE 1: 2-MONTH DATA WAS USED

Furnace variables have been tabulated in Table 11.1, while the correlation matrix is presented in Figure 11.1. In this work, hot metal temperature (HMT) and silicon content of hot metal (Si) are the objectives. The rest of the parameters are design variables. For a smoother downstream operation during refining, it is important to check up on the hot metal tapped from a blast furnace. The silicon content of the hot metal must be kept at optimum levels (low). Hot metal temperature must be high enough to maintain fluidity, but a higher hot metal temperature will enhance the silicon content of the hot metal. Thus, both objectives are conflicting. In the

TABLE 11.1

Furnace 1 Variables and Objectives

Process Variables and Objectives	Min.	Max.
Hot blast temperature (HBT)	1171.819	1228.371
Raceway adiabatic flame temperature (RAFT)	2154.751	2334.598
Cold blast flow(CBF)	280000	351667.3
Hot blast pressure (HBP)	3.3	4.321077
Top pressure (TP)	1.8	2.3865
Total_K (K)	3.127824	3.977886
Uptake temp (Uptake_temp)	85.95206	177.237
Oxygen enrichment (O2_ENRICH)	2.92825	7.156007
Coke rate(COKR_RT)	387.3612	601.9096
Hot metal temperature (HMT)	1409	1585
Silicon in hot metal (SI)	0.1	1.94

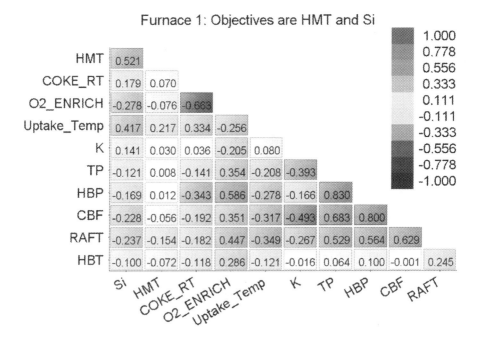

FIGURE 11.1   Correlation matrix for furnace 1.

following sections, we present a set of approaches that will be helpful for analyzing the blast furnace process variables, and their correlations with the objectives.

From Figure 11.1, one can observe that some correlations exist between design variables and objectives. This data is highly noisy, just like any other data set from an industrial furnace.

11.3.1 SUPERVISED MACHINE LEARNING IN RAPIDMINER

Supervised machine learning models were developed for the hot metal temperature (HMT) and silicon in hot metal (Si) in the RapidMiner software. Table 11.2 provides error metrics for hot metal temperature (HMT). Table 11.3 provides error metrics for silicon in hot metal (Si). Figure 11.2 shows the comparison between actual silicon content in hot metal and model predictions.

One can observe that the relative error for silicon in hot metal (Si) is quite high for all the models. This is possible as these data sets are quite noisy. Silicon content fluctuates a lot, so such predictions are expected. Still, there is not much time wasted as training/development of these models is quick. RapidMiner software is free for data sets up to 10,000 rows and no programming is involved; just a few clicks of buttons, and models are trained.

Supervised ML models for hot metal temperature (HMT) are acceptable. Thus, we will work with other approaches on this data set for determining various correlations. For industrial data, one needs to work on several approaches as the correlations are highly non-linear. This data was processed and was clean from

TABLE 11.2

Furnace 1: Error Metrics for Hot Metal Temperature (HMT) Models Developed in RapidMiner

Model	Relative Error (%)	SD (%)	Total Time(s)
Generalized Linear Model	1.131	0.0592	1.243
Deep Learning	1.161	0.0861	1.529
Decision Tree	1.170	0.0708	0.276
Random Forest	1.137	0.0724	2.967
Gradient Boosted Trees	1.155	0.0731	15.096
Support Vector Machine	1.160	0.1152	7.625

TABLE 11.3

Furnace 1: Error Metrics for Silicon Content in Hot Metal (Si) Models Developed in RapidMiner

Model	Relative Error	SD	Total Time (s)
Generalized Linear Model	0.163	0.014	0.199
Deep Learning	0.165	0.009	0.959
Decision Tree	0.167	0.010	0.148
Random Forest	0.155	0.012	3.218
Gradient Boosted Trees	0.164	0.010	10.903
Support Vector Machine	0.169	0.017	210.740

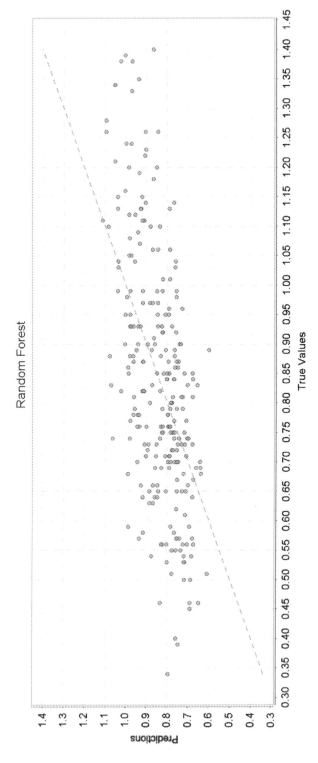

FIGURE 11.2 Furnace 1 silicon: Actual vs. predicted values for model developed through random forest algorithm in RapidMiner.

outliers, AI/ML-based models for one objective are acceptable, while for the other objective the model accuracy is not acceptable.

11.3.2 Unsupervised Machine Learning in modeFRONTIER

For SOM analysis, X-Dimension is 11 and Y-Dimension = 12; thus, total map units are equal to 132. The neighbors function is Gaussian. For reducing error, data were scaled through logistic scaling. The quantization error is 0.065 while the topological error is 0.027. Thus, both errors are acceptable as the data are highly noisy. Figure 11.3 shows the SOM component plots for all the variables and objectives included in Table 11.1.

From Figure 11.3, one can observe that silicon in hot metal (Si) is grouped with uptake temperature, while hot metal temperature (HMT) is grouped with hot blast temperature (HBT). Both correlations can be verified from literature. Additionally, oxygen enrichment (O2_ENRICH) and coke rate (COKE_RT) are grouped together, and this relation can also be verified from literature. Thus, from this data set, one can draw meaningful correlations that can be verified from literature. Even the units for hot blast pressure (HBP) and top pressure (TP) are close to each other. In blast furnace operations, sometimes these variables are combined together, and pressure drop is calculated as a difference between HBP and TP, just like K is the average of four (4) values of K (K1, K2, K3, and K4). RAFT is close to oxygen enrichment and coke rate. It's a known fact that oxygen enrichment will lead to burning of coke, which will result in an increase in RAFT. Figure 11.4 shows the

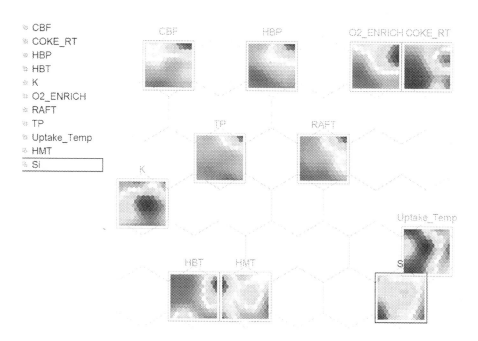

FIGURE 11.3 Furnace 1: SOM component map for various variables and objectives.

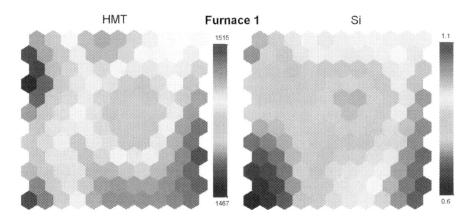

FIGURE 11.4 Furnace 1: SOM analysis for hot metal temperature (HMT) and silicon in hot metal (Si).

distribution of hot metal temperature (HMT) and silicon in hot metal (Si) on the SOM space for Furnace 1.

In Figure 11.4, one can observe that an increase in silicon content is directly related with hot metal temperature. A few regions exist on the SOM map that can be helpful for the operator, like a few hexagonal units just below the HMT and Si labels on Figure 11.4. The temperature for these units is slightly below the maximum values if one compares the color with the color bar. Similarly, the silicon content of these few units is in the medium range, which can be observed by comparing the color over these units and the color bar for silicon. Variables associated with these regions can be helpful for the operators in determining optimum furnace process parameters.

11.3.3 Multi-Objective Optimization in modeFRONTIER

11.3.3.1 Variable Screening Through ANOVA

Variable screening was performed in modeFRONTIER software using smoothening spline ANOVA algorithm. Contribution indexes for process parameters are listed in Table 11.4. One can observe that for HMT, error values are acceptable, but the R-squared value is quite low at 0.098, whereas for silicon in hot metal (Si), the R-squared value is slightly better at 0.412 but error values are still a bit high.

11.3.3.2 Model Development by Supervised Machine Learning

Supervised ML-based predictive models were developed for hot metal temperature (HMT) and silicon in hot metal (Si) using the K-nearest neighbor algorithm. A comparison between actual values and K-NN model predictions for silicon in hot metal (Si) is shown in Figure 11.5, while a similar plot for hot metal temperature (HMT) is shown in Figure 11.6.

TABLE 11.4

Contribution Index of Various Process Parameters for Furnace 1

	HBT	RAFT	CBF	HBP	TP	K	Uptake_Temp	O2_ENRICH	COKE_RT	Mean Absolute Error	Mean Relative Error	Mean Normalized Error	R-squared
HMT	0.015	0.08	0.278	0.109	−0.026	−0.087	0.399	0.26	−0.027	17.203	0.011	0.1	0.098
Si	0.018	−0.114	0.577	−0.485	0.245	−0.101	0.387	0.523	−0.049	0.135	0.174	0.119	0.412

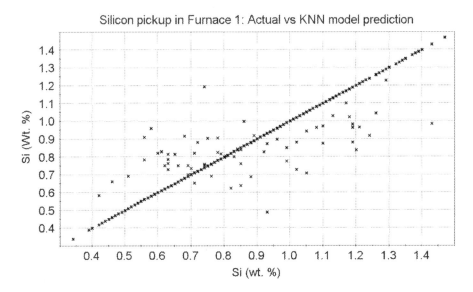

FIGURE 11.5   Furnace 1 Silicon: Actual vs. predicted values (K-NN model).

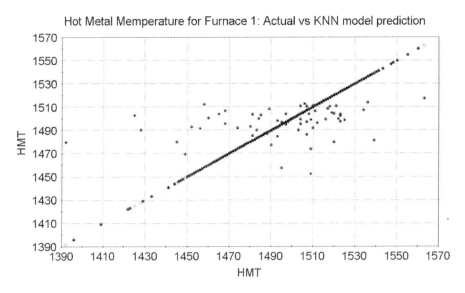

FIGURE 11.6   Furnace 1 HMT: Actual vs. predicted values (K-NN model).

From Figures 11.5 and 11.6, one can observe that model predictions for a lot of values lie on the 45-degree line. In some cases, predictions are near the 45-degree line, while in some cases the model predictions are far apart from the actual values. As per our experience in dealing with blast furnace data of various furnaces around the world (Jha et al. 2014, Mahanta et al. 2022), these predictions are acceptable. We have reported similar trends in these and other published works.

11.3.3.3   Multi-Objective Optimization Through Genetic Algorithm

After developing models, we proceeded towards multi-objective optimization. Workflow was developed in the modeFRONTIER software (Figure 11.7). A number of algorithms listed in Chapter 2 were used for this work.

The problem was formulated as follows:

- Maximize hot metal temperature (HMT) and
- Minimize silicon in hot metal (Si)

In Figure 11.7, one can observe that there are constraints on both hot metal temperature (HMT) and silicon in hot metal (Si). Figures 11.8 shows the Pareto-frontier obtained after solving the multi-objective problem formulated for Furnace 1. An operator can define these constraints as per their knowledge and experience with a particular furnace (Figures 11.7).

11.4   FURNACE 2: 1-MONTH DATA WAS USED

One-month data from another operating furnace was used for this study. Data was preprocessed, as blast furnace data are quite noisy and there are many outliers. This is mainly because the data points were collected from the real-world environment around a blast furnace, due to which some of these errors are unavoidable.

11.4.1   ANALYSIS OF DATA WITH RESPECT TO OPERATING PARAMETERS AND OBJECTIVES

In Table 11.5, a total of sixteen (16) process parameters are listed. These process parameters were identified to affect the objectives and the constraint mentioned later in the text. These process parameters were further modified, and the updated names have been listed in Table 11.6.

The available input variable data points were used for training mutually conflicting objectives:

- Hot metal temperature (°C)
- Total input carbon (Kg/THM): Obtained by adding coke carbon and PCI carbon.
- Silicon content (%) in the hot metal

Variable screening was performed in the modeFRONTIER software using the smoothening spline ANOVA algorithm. The contribution indexes for process parameters are listed in Table 11.6.

From Table 11.6, one can observe that the R-squared value is acceptable for both hot metal temperature (HMT) and silicon in hot metal (Si). Error metrics obtained through statistical analysis are low and acceptable. In this case everything looks fine; a lot of mixed correlations are expected. But, the size of the data set is quite small as we are analyzing data from just one month. Additionally, a number of operating parameters are 16, which adds to the complexity of the problem.

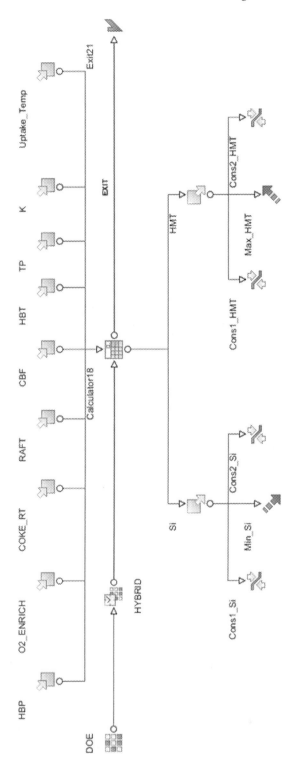

FIGURE 11.7 Furnace 1: Workflow for performing muti-objective optimization.

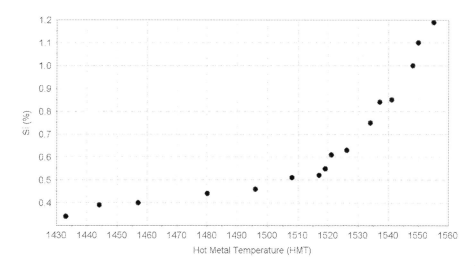

FIGURE 11.8 Furnace 1: Optimized values for hot metal temperature (HMT) and silicon in hot metal (Si).

TABLE 11.5
Furnace 2: Operating Parameters

Sl. No.	Variable	Unit
1	Iron ore	mt/thm
2	Sinter	mt/thm
3	Pellet	mt/thm
4	Coke	Kg/THM
5	Pulverised coal injection (PCI)	Kg/THM
6	Oxygen enrichment	%
7	Blast humidification	g/Nm^3
8	Average blast volume	Nm^3/min
9	Hot blast pressure	Kg/cm^2
10	Hot blast temperature	°C
11	Top pressure	Kg/cm^2
12	Raceway adiabatic flame temperature (RAFT)	°C
13	Total tap time	Hrs.
14	Hot metal flow rate	MT/minute
15	K1	Ratio
16	K2	Ratio

Thus, we analyzed the data using the concept of self-organizing maps (SOMs). The SOM algorithm can be used for a small database of up to 25–30 rows, which is just the case here. Additionally, the SOM algorithm can be used for determination of complex pattern in multi-dimensional data.

TABLE 11.6

Furnace 2: Modified Operating Parameters and Their Contribution Index with Respect to the Objectives

Sl. No.	Variable	Unit	HMT	HM_Si (Silicon Content)
1	Total Fe (Ore + Pellet) (Total_Fe_Ore_Pellet)	mt/THM	0.111	−0.015
2	Fe in sinter (Fe_Sint)	mt/THM	0.095	−0.022
3	FeO in sinter (FeO_Sint)	mt/THM	5.3E-4	0.143
4	Coke carbon (Coke _Carbon)	Kg/THM	−0.015	−0.112
5	PCI carbon (PCI_Carbon)	Kg/THM	0.219	0.692
6	Oxygen enrichment (Oxy_Enrh)	%	−0.098	0.193
7	Blast humidification (Blst_Humid)	g/Nm^3	0.358	0.405
8	Average blast volume (Av_Bl_Vol)	Nm^3/min	−0.073	0.034
9	Hot blast pressure (Hot_Blst_Press)	Kg/cm^2	0.037	0.109
10	Hot blast temperature (Hot_Blst_Temp)	°C	−0.075	0.069
11	Top pressure (Top_Press)	Kg/cm^2	−0.011	0.025
12	Raceway adiabatic flame temperature (RAFT)	°C	−0.006	−0.093
13	Total tap time (Tot_Tap_Time)	Hrs.	0.3163	−0.189
14	Hot metal flow rate (Hot_Metal_Flow_Rate)	MT/minute	0.213	−0.238
15	K1	Ratio	0.136	−6.96E-4
16	K2	Ratio	−0.017	−9.09E-4
Mean Absolute Error			2.252	0.023
Mean Relative Error			0.002	0.053
Mean Normalized Error			0.063	0.065
R-squared			0.847	0.773

11.4.2 UNSUPERVISED MACHINE LEARNING: SELF-ORGANIZING MAPS

In SOM analysis, X-Dimension = 5, Y-Dimension = 5; thus, total map units are equal to 25. The neighbors function was chosen as Gaussian. For reducing error, data were scaled during the SOM analysis and the approach used in mode-FRONTIER is called logistic scaling. The quantization error is 0.141, which is on the higher side. Thus, one must be careful while using these plots as a predictive tool. However, the topological error was calculated as 0.00001, which is quite low. This means that the topology of the data was perfectly preserved. Thus, the SOM model can be used for determining various patterns in the data set.

The SOM component plot for Furnace 2 is plotted in Figure 11.9. It must be noted that there are 16 process parameters and 3 objectives. All of these 19 columns have been plotted on the SOM component plot. Readers must refer to Table 11.6, as the numbers in Figure 11.9 are the serial numbers from Table 11.6. Hot metal temperature (HMT) and silicon in hot metal (HM_Si) units are adjacent to each

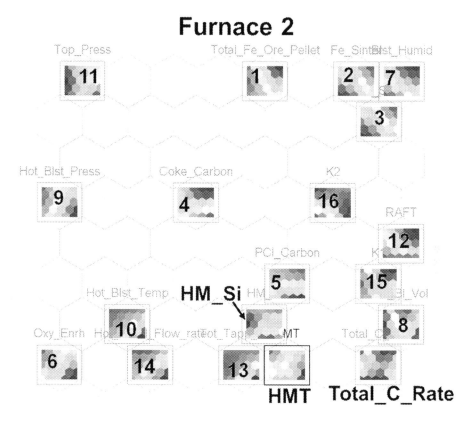

FIGURE 11.9 Furnace 2: SOM component plot.

other and grouped with total tapping time and PCI_carbon. Another objective was total carbon rate, and this unit is close to the hot metal temperature (HMT). The unit for total carbon rate is also grouped with average blast volume. Units marked 1, 2 and 3 represent the iron content and these units are close to each other. Correlations between the units grouped together on the SOM component map can be verified from literature.

Thus, even for a small database of a month, the SOM algorithm was able to pick lots of important correlations. This data set is small, and too many variables are there, but such a case can arise in the industry, especially if there are some problems with the furnace. Rather than analyzing a large data set and developing complex models, a user can use the SOM analysis to study the pattern. An operator can utilize their expertise and come up with a solution as they know the furnace behavior over time.

For smooth running of a blast furnace, it is extremely important that day-to-day operations are routinely monitored. Approaches like the SOM analysis can be quite helpful for analyzing patterns within a high-dimensional data set. In this work, we have demonstrated that even one month of data can be efficiently utilized.

In Figure 11.10, we have plotted the three objectives, along with three operating parameters. On the objectives' plots, there are units highlighted with a number.

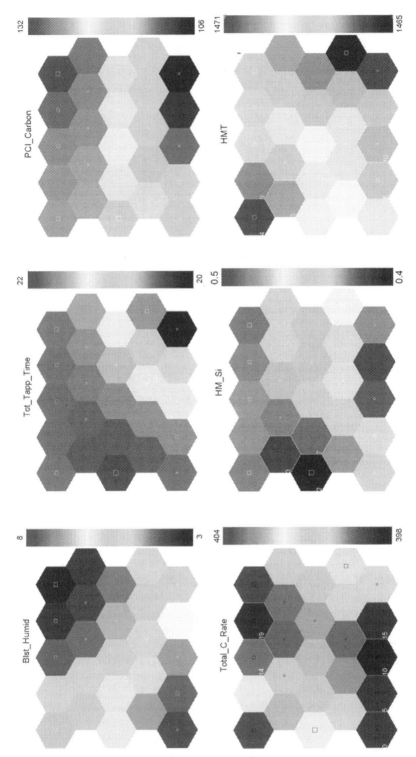

FIGURE 11.10 Furnace 2: SOM plots for the objectives, along with a few relevant operation parameters.

These are the best units identified in the present analysis of these objectives. The three operating parameters on these figures are the ones that were identified through SOM component plot analysis and through variable screening through the ANOVA algorithm. One can observe that one needs to find a compromise between the three objectives for improving upon performance on all three fronts. This is what we mention as the concept of Pareto-optimality in Chapter 3.

The unit with the highest HMT is associated with a higher total carbon rate. The two units around that unit can be considered for further examination if one wants to check upon the operating parameters that will provide them with optimized values for total carbon rate, along with hot metal temperature (HMT) and silicon content in hot metal. Since there are 16 operating parameters, it is advised to arrange more data prior to reaching a conclusion. Our purpose of this analysis was to demonstrate ways of efficiently analyzing data from just one month from a highly non-linear process of an iron-making blast furnace.

11.5 FURNACE 3: 1-YEAR DATA WAS USED FOR THIS ANALYSIS

11.5.1 Data Processing

Operating parameters for Furnace 3 have been tabulated in Table 11.7. Data was screened for outliers and preprocessed. Four (4) objectives were identified, CO_2 content in the exit gas, productivity of the furnace, hot metal temperature, and silicon content in the hot metal.

Figure 11.11 shows the correlation matrix over the data obtained from Furnace 3. Productivity and CO_2 content in the exit gas are strongly correlated. Another strong correlation is between top pressure and average blast pressure. Apart from these, most of the correlations are weak. Statistical analysis was performed on this data set using a smoothing spline ANOVA algorithm in modeFRONTIER for determining the effect of operating parameters of a furnace on the objectives/constraints.

TABLE 11.7

Furnace 3 Operating Parameters

Sl. No.	Input Parameters	Unit
1	Hot Blast Temperature (HBT)	°C
2	Blast Humidification (Steam Rate (StRt))	gm/Nm^3
3	Average Blast Pressure (AvBlPress)	Kg/cm^2
4	Average Blast Volume (AvBlVol)	Nm^3/THM
5	Top Pressure (Top Press)	Kg/cm^2
6	Coke Ash	%
7	Coke Moisture (Coke Moist)	%
8	Sinter Ore Ratio	–
9	Carbon Rate (C Rate)	Kg/THM

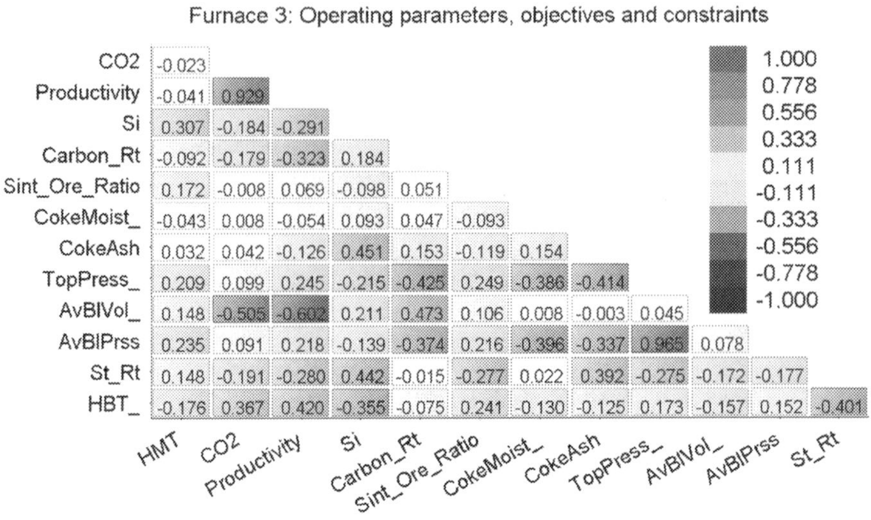

Furnace 3: Operating parameters, objectives and constraints

	HMT	CO2	Productivity	Si	Carbon_Rt	Sint_Ore_Ratio	CokeMoist_	CokeAsh	TopPress_	AvBlVol_	AvBlPrss	St_Rt
CO2	-0.023											
Productivity	-0.041	0.929										
Si	0.307	-0.184	-0.291									
Carbon_Rt	-0.092	-0.179	-0.323	0.184								
Sint_Ore_Ratio	0.172	-0.008	0.069	-0.098	0.051							
CokeMoist_	-0.043	0.008	-0.054	0.093	0.047	-0.093						
CokeAsh	0.032	0.042	-0.126	0.451	0.153	-0.119	0.154					
TopPress_	0.209	0.099	0.245	-0.215	-0.425	0.249	-0.386	-0.414				
AvBlVol_	0.148	-0.505	-0.602	0.211	0.473	0.106	0.008	-0.003	0.045			
AvBlPrss	0.235	0.091	0.218	-0.139	-0.374	0.216	-0.396	-0.337	0.965	0.078		
St_Rt	0.148	-0.191	-0.280	0.442	-0.015	-0.277	0.022	0.392	-0.275	-0.172	-0.177	
HBT_	-0.176	0.367	0.420	-0.355	-0.075	0.241	-0.130	-0.125	0.173	-0.157	0.152	-0.401

Legend: 1.000, 0.778, 0.556, 0.333, 0.111, -0.111, -0.333, -0.556, -0.778, -1.000

FIGURE 11.11 Furnace 3: Correlation matrix.

11.5.1.1 Variable Screening

Variable screening was performed in the modeFRONTIER software using a smoothing spline ANOVA algorithm. Contribution indices are listed in Table 11.8.

In Table 11.8, it can be observed that an increase in average blast volume leads to an increase in both CO_2 and productivity. An increase in average blast pressure will lead to an increase in hot metal temperature, while an increase in coke ash content will lead to an increase in silicon content of hot metal. These correlations have been observed by operators, but it can be observed that most of the other correlations are quite weak, which is expected for such noisy databases. It is always recommended to preprocess data and use statistical tools for analyzing correlations within the data set. Thus, a data scientist can analyze those patterns and look at the literature or consult a furnace operator for their views. Some of the patterns can be directly understood by an operator, while some are too complex to quantify due to multiple operating parameters.

11.5.2 Unsupervised Machine Learning: Self-Organizing Maps

The data set for Furnace 3 was analyzed through the concept of self-organizing maps (SOMs). In SOM analysis, X-Dimension = 9, Y-Dimension = 11; thus, the total map units are equal to 99. The neighbors function was chosen as Gaussian. For reducing error, data were scaled during the SOM analysis and the approach used in the modeFRONTIER is called logistic scaling. The quantization error is 0.100, which is on the higher side. The topological error was calculated as 0.022, which is quite low. This means that the topology of the data was perfectly preserved. Thus, the SOM model can be used for determining various patterns in the data set. Figure 11.12 shows the SOM component plot for operating parameters and targeted objectives and constraints for Furnace 3.

TABLE 11.8

Furnace 3: Variable Screening through Smoothing Spline ANOVA Algorithm in modeFRONTIER

Sl. No.	Input Parameters	CO_2	Productivity	Hot Metal Temperature	Hot Metal Silicon
1	Hot Blast Temperature (HBT)	0.136	0.121	0.150	0.180
2	Blast Humidification (Steam Rate (StRt))	0.095	0.130	0.142	0.277
3	Average Blast Pressure (AvBlPress)	0.169	0.177	0.370	–0.104
4	Average Blast Volume (AvBlVol)	0.654	0.692	0.127	0.101
5	Top Pressure (Top Press)	–0.037	0.058	–0.083	0.133
6	Coke Ash	0.017	–0.015	0.009	0.391
7	Coke Moisture (Coke Moist)	0.004	–0.010	–0.001	0.005
8	Sinter Ore Ratio	0.002	–0.002	0.204	–0.011
9	Carbon Rate (C Rate)	–0.039	–0.036	0.091	0.026
Mean Absolute Error		0.035	0.042	4.445	0.049
Mean Relative Error		0.033	0.028	0.003	0.087
Mean Normalized Error		0.092	0.081	0.143	0.125
R–squared		0.548	0.658	0.208	0.424

In Figure 11.12, one can observe that the silicon content in hot metal and steam rate is adjacent to each other on the component plot. The relation between blast humidification and hot metal silicon was also observed for Furnace 2. These are different furnaces, but the final product for both furnaces is the same: hot metal, slag, and exit gas. One must not expect similar correlations or correlations of the same intensity for different furnaces. The reason is that each furnace has different sources for raw material, located in different geographical locations with a different weather pattern. Blast furnace iron making is an open process. Figure 11.13 shows the distribution of the two objectives, two constraints, and two operating parameters.

In Figure 11.13, one can observe that productivity and CO_2 are conflicting in nature. A region of high CO_2 and productivity coincide on the SOM plot. Thus, one needs to find a balance between both of these main objectives. Thus, we perform multi-objective optimization using several concepts of artificial intelligence for developing surrogate models as well as optimizing the objectives and keeping constraints within prescribed bounds.

11.5.3 Multi-Objective Problem Formulation

The available input variable data points were used for training mutually conflicting objectives:

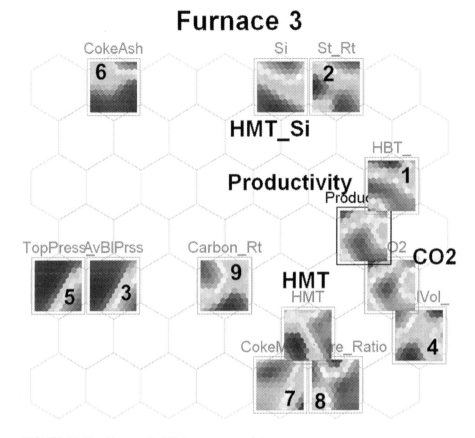

FIGURE 11.12   Furnace 3: SOM component plot.

- Productivity ((THM/Day)/Working Volume) and
- Net CO_2 ((T/Day)/Working Volume) in exit gas stream

The objectives for Furnace 3 are different from Furnaces 1 and 2. For Furnace 3, hot metal temperature (HMT) and silicon in hot metal (Si) are the constraints, whereas for Furnace 1 and 2, these were part of the main objective.

There was a constraint on silicon content (weight%) in the hot metal and hot metal temperature. The silicon content (%) in the hot metal often shows erratic fluctuations, reflecting lower zone inefficiencies and thus affecting the quality of hot metal. Therefore, attempts were made to contain it within some prescribed bounds. In the present case, it was decided to keep the silicon content in the hot metal in the range of 0.4–0.8%. Further, this range was subdivided and now silicon content in hot metal was constrained in three bounds:

- Low silicon (0.40–0.55%)
- Medium silicon (0.55–0.70%)
- High silicon (0.70–0.80%)

FIGURE 11.13 Furnace 3: SOM plots for the objectives, constraints, and operating parameters.

In this case, the goals set are conflicting. In other words, if we try to improve one of them, it will result in a deterioration of the other. Thus, this problem is well suited for multi-objective optimization.

11.5.3.1 Case 1: Two Objectives and One Constraint

The multi-objective optimization problem was formulated as stated in the following text, while the workflow developed in modeFRONTIER is shown in Figure 11.14.

"Maximizing productivity and minimizing net CO_2 in the exit gas stream subject to a constraint on the silicon content in the hot metal".

In Figures 11.15–11.17, we obtained a well-defined Pareto set for all the cases. Legends in the figure correspond to the notations used for a particular approach or algorithm. All of these notations have been mentioned in Chapter 3.

11.5.3.2 Case 2: Two Objectives and Two Constraints

Multi-objective optimization problem was formulated as mentioned below.

"Maximizing productivity and minimizing net CO_2 in the exit gas stream subject to a constraint on the silicon content in the hot metal and hot metal temperature".

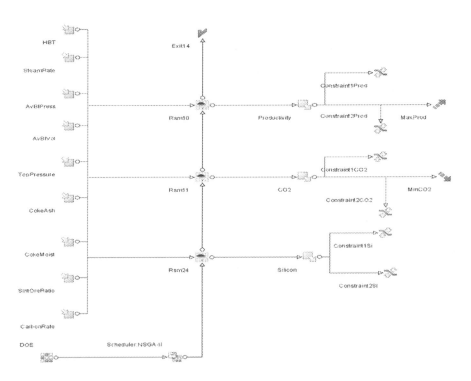

FIGURE 11.14 Furnace 3: Workflow in modeFRONTIER.

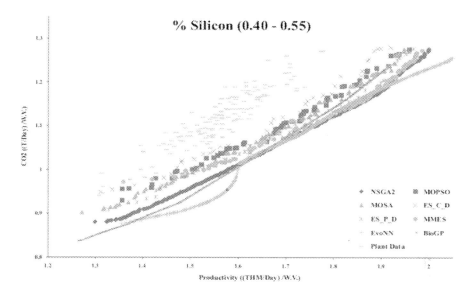

FIGURE 11.15 Furnace 3: Maximize productivity, minimize CO_2 in exit gas. Silicon in hot metal: 0.40–0.55.

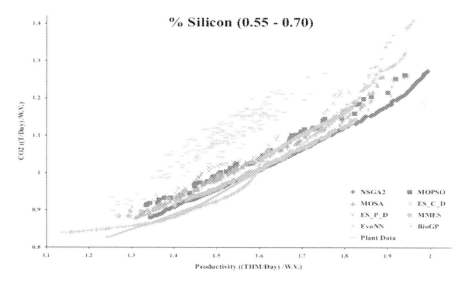

FIGURE 11.16 Maximize productivity, minimize CO_2 in exit gas. Silicon in hot metal: 0.55–0.70.

In case 2, hot metal temperature (HMT) was varied between 1,425 to 1,460°C. Silicon content in the hot metal was divided in three regimes: Si4055 (0.40 to 0.55 weight% silicon), Si5570 (0.55 to 0.70 weight% silicon) and Si7080 (0.70 to 0.80 weight% silicon). Pareto-frontier obtained through evolutionary neural network (EvoNN) and bi-objective genetic programming (BioGP) has been shown in Figures 11.18 and 11.19 respectively.

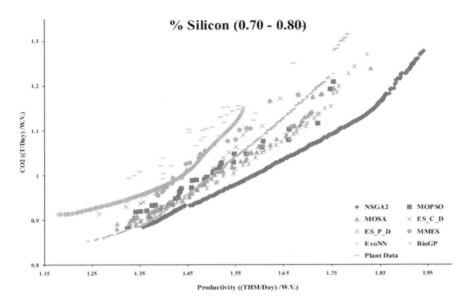

FIGURE 11.17 Maximize productivity, minimize CO_2 in exit gas. Silicon in hot metal: 0.70–0.80.

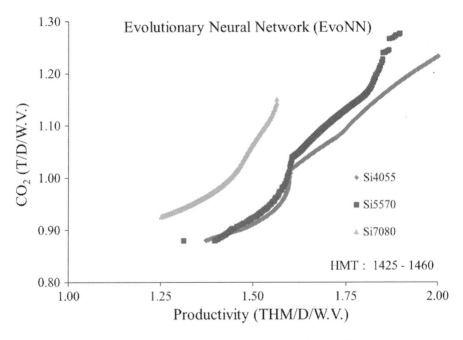

FIGURE 11.18 Maximize productivity, minimize CO_2 in exit gas. Hot metal temperature: 1425–1560°C. Silicon in hot metal: 0.40–0.80 weight%.

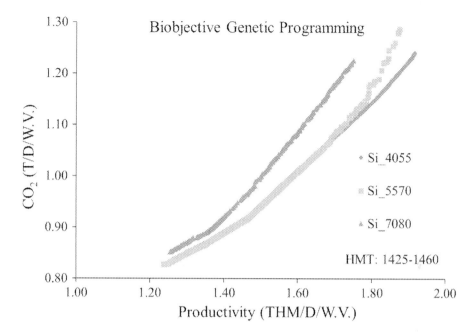

FIGURE 11.19 Maximize productivity, minimize CO_2 in exit gas. Hot metal temperature: 1425–1560°C. Silicon in hot metal: 0.40–0.80 weight%.

The developed models were able to capture the basic trends shown in the actual plant data. The various interrelationships shown by our model can be verified through literature.

The following conclusions can be drawn from the optimization results for Furnace 3:

1. A productivity boost of about 10–12%
2. A reduction in the CO_2 content of the outlet gas by approximately 8–10%
3. A decrease in net carbon content by approximately 8–10%

12 Case Study #12: Industrial Furnaces II: Development of GUI/APP to Determine Additions in LD Steelmaking Furnace

12.1 INTRODUCTION: LD STEELMAKING

In this chapter, we use data from an operating basic oxygen furnace (BOF). It will also be referred as LD furnace throughout the chapter. The LD furnace has been described in detail in Chapter 2. Thus, in this chapter, we will discuss a few parameters of that furnace from an integrated steel plant.

Furnace capacity is 180 tons. Blowing operation, or the process of lancing of oxygen, is performed for an average of around 16 to 22 minutes. During blowing, several exothermic reactions take place that raise the temperature of the bath. Additionally, during the blowing operation several additions are performed in achieving desired chemistry of the steel. An operator works on optimizing operating cost, productivity, and quality of steel. One of the major challenges an operator faces is maintaining the temperature of the bath. Almost all the steel plants use some sort of automation model to help the operator. Automation models provide a guideline on the basis of inputs provided by the operator. These guidelines are in the form of predicted values of additions to be made by an operator to achieve optimum temperature control. Automation models for BOF are based on preset calculations. This model takes into account the operating parameters and usually provides predictions on the basis of heat and mass balance. Heat and mass balance are performed to achieve desired chemistry and aim temperature. Such calculations are performed through the model prior to each blowing cycle.

Now, the most important part is whether the operator will use these predictions from the automation model, or will they go by their own intuition? Intuition of an operator can also be considered a prediction method. An experienced operator can have a better understanding of a system from years of experience through working in that surrounding.

Though these models are helpful, it is still based on any pre-set conditions that the software developer was introduced to. In the previous chapter on blast furnace

DOI: 10.1201/9781003167372-12

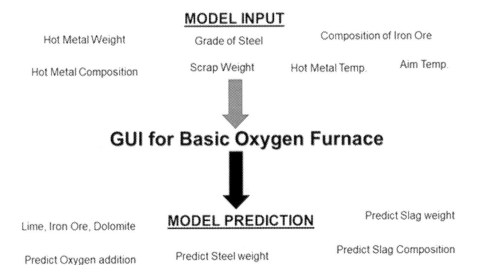

FIGURE 12.1 Workflow for GUI/APP development.

iron making, we explained it in detail. In materials/alloy design, researchers across the globe can acquire same raw materials from the same vendor and can perform experiments and characterization using the same techniques and equipment. But, in process metallurgy, it is impossible to even imagine such a thing. An integrated plant is a size of a small city, and these furnaces are located in different geographical locations, use different raw materials from places with completely different weather pattern.

Thus, the main purpose of the BOF operator is to achieve desired chemistry at a defined temperature. These automation models can be helpful, but final decision is taken by the operator regarding additions, time of blowing, etc.

In this chapter, we have presented material balance equations that will be helpful for a reader in developing an automation model for a BOF/LD furnace. The model was embedded in the form of a graphical user interface (GUI) or as a MATLAB® APP. The workflow is shown in Figure 12.1.

12.2   DEVELOPMENT OF GUI: FORMULATION OF ANALYTICAL MODEL

MATLAB programming language was used for developing this GUI.

12.2.1   CALCULATIONS: ASSUMPTIONS

Data for one year was analyzed for this present BOF furnace. The most common basicity value was 3.3 and slag FeO was around 20%. Thus, these values were fixed during this study, though one can choose a different basicity in our GUI/APP.

Fumes in output were assumed to be less than 2 tons.

- Iron-Ore Composition
  - % Fe2O3 in ore = 93%
  - % Fe in ore = 65%
  - % SiO2 in ore = 3%
- Scrap:
  - Scrap Fe % = 98%
  - Slag FeO = 20%

12.2.2 MATERIAL BALANCE

$$= >\ W_{HM} + W_{SCRAP} + W_{ORE} + W_{LIME} = W_{STEEL} + W_{SLAG} + W_{FUMES}$$

Input

$$=> W_{HM} + W_{SCRAP} + W_{ORE} + \left\{ \frac{\% Si\ hm * W\ hm}{100} * 2.14) + \frac{\% SiO2\ ore * W\ ore}{100} \right\} * B +$$

$$\left\{ \frac{0.5}{5500}(\% Mn\ Hm * W\ Hm - \% Mn\ Steel * W\ Steel) + \frac{1.25}{3100}(\% P\ Hm * W_{Hm} - \% P\ Steel * W\right.$$

$$\text{steel}) + \frac{\% Si\ Hm * W\ Hm}{1200} * 0.533 + \frac{\% C\ Hm * W\ Hm - \% C\ Steel * W\ Steel}{1200} * (x + 0.5(1 - x) -$$

$$\frac{\% Fe2O3\ Ore * W\ Ore}{100 * 159.7} * 1.5 + \left[\left\{ \left(\frac{\% Si\ hm * W\ hm}{100} * 2.14 + \frac{\% SiO2 * W\ ore}{100} \right) * (B + 1) \right\} \right.$$

$$\left(\frac{\% Mn\ hm * Whm - \% MnSteel * WSteel}{100} * 1.291 \right) + \left(\frac{\% P\ hm * Whm - \% PSteel * WSteel}{100} * 2.29 \right) \right] *$$

$$\left. \frac{5}{4 * 5 * 2 * 71.85} \right\} * 32$$

Output

$$W_{Steel} + \left\{ \left[\left(\frac{\% Si\ hm * W\ hm}{100} * 2.14 \right) + \frac{\% SiO2\ ore * W\ ore}{100} \right] * (B+1) \right.$$

$$+ \left(\frac{\% Mn\ hm * Whm - \% Mn\ Steel * WSteel}{100} * 1.291 \right) \left(\frac{\% P\ hm * Whm - \% P\ Steel * WSteel}{100} * 2.29 \right) \right\} * \frac{5}{4}$$

$$+ \frac{\% C\ hm * Whm - \% C\ Steel * WSteel}{100} * \left(x * \frac{1.25}{12} + (1-x) * \frac{28}{12} \right)$$

$$=> W_{HM} + W_{SCRAP} + W_{ORE} + \left\{ \left(\frac{\% Si\ hm * W\ hm}{100} * 2.14 \right) + \frac{\% SiO2\ ore * W\ ore}{100} \right\} * B +$$

$$\left\{ \frac{0.5}{5500}(\% Mn\ Hm * W\ Hm - \% Mn\ Steel * W\ Steel) + \frac{1.25}{3100}(\% P\ Hm * W_{Hm} - \% P\ Steel * \right.$$

$$\text{W steel}) + \frac{\% Si\ Hm * W\ Hm}{1200} * 0.533 + 1.21$$

Final equation for steel and ore prediction:

$$= > \text{W steel } [1 - 0.01251 * \text{Mn \% steel} - 0.014447 * \%\text{P steel}$$
$$- 0.01 * \text{C\% Steel}]$$
$$-\text{W ore } [1 - (\% \text{ SiO2 Ore}) * (0.019433 * B + 0.019433)$$
$$- 0.03006 * \% \text{ Fe}_2\text{O}_3]$$
$$= \text{W hm} + \text{W Scrap} - \%\text{Si Hm} * \text{W Hm} * (0.011345 + 0.0041587 * B)$$
$$- 0.010576 * \% \text{ Mn Hm} * \text{W Hm} - 0.0144469 * \%\text{P hm} * \text{W Hm}$$
$$- 0.01 * \%\text{C Hm} * \text{W Hm}$$

$$(12.1)$$

For equation 12.2: Fe balance:

$$\Rightarrow \left(\frac{\% Fe\ Hm * Wt\ Hm}{100} \right) + \left(\frac{\% Fe\ Scrap * Wt\ Scrap}{100} \right) + \left(\frac{\% Fe\ Ore * Wt\ Ore}{100} \right)$$
$$= \left(\frac{\% Fe\ Slag * Wt\ Slag}{100} \right) * \frac{55.85}{71.85} + \left(\frac{\% Fe\ Steel * Wt\ Steel}{100} \right) + \frac{53.4}{100} * \text{fumes}$$

Final equation 12.2 for prediction of W steel and W ore:

$$= > 0.01 (\%\text{Fe}_{hm} * \text{Wt}_{hm}) + 0.01 (\%\text{Fe}_{scrap} * \text{Wt}_{scrap}) - 0.0041516$$
$$* \{(B + 1) * (\%\text{Si}_{hm} * \text{Wt}_{hm})\}$$
$$- 0.0025 * (\%\text{Mn}_{hm} * \text{Wt}_{hm}) - 0.0044426 * (\%\text{P}_{hm} * \text{Wt}_{hm}) + 1.211$$

$$= \text{Wt}_{steel} * (0.01 * \%\text{Fe}_{steel} - 0.0025 * \%\text{Mn}_{steel} - 0.00444 * \%\text{P}_{steel})$$
$$- \text{Wt}_{ore} * (0.01 * \%\text{Fe}_{ore} - 0.00415 * (B + 1) * \%\text{SiO}_2)$$

$$(12.2)$$

By solving equations 12.1 and 12.2, we get the prediction of W steel and W ore.
X = Weight of steel
Y = Weight of iron-ore needed

Calculating lime and dolomite weight:

$$\text{SiO}_2\text{Slag} = 2.14 * B * W_{Hm} * (\% \text{ Si Hm}/100)$$
$$\text{CaO slag} = \text{SiO}_2\_\text{slag} * B$$
$$\text{MgO slag} = 0.48 * \text{SiO}_2\text{slag}$$

Assuming lime gives 96% CaO and 2% MgO
Assuming dolo gives 56% CaO and 40% MgO

We get a set of simultaneous equations:

$$0.96(x) + 0.56(y) = CaO \ slag \qquad (12.3)$$

$$0.02(x) + 0.40(y) = MgO \ slag \qquad (12.4)$$

By solving equations 12.3 and 12.4, we can determine the lime and dolomite addition.

Here, X = **lime needed**

Y = **dolomite needed**

For slag calculation:

$$SiO_2 \ Slag = 0.01 * ((\%Si \ Hm * W \ Hm * 2.14) + (\%SiO_2 * W \ ore))$$

$$CaO \ Slag = SiO_2 \ Slag * Basicity$$

$$Mno \ Slag = 0.01 * ((W \ Hm * \%Mn \ Hm) - (\%Mn \ Steel * W \ Hm * 1.291))$$

$$P_2O_5 \ Slag = 0.01 * ((W \ Hm * \%P \ Hm) - (2.29 * \%P \ Steel * W \ Steel))$$

$$\textbf{W Slag} = (SiO_2 \ slag + CaO \ slag + MnO \ slag + P_2O_5 \ Slag)$$

$$FeO \ slag = 0.2 * W \ Slag$$

For oxygen-required calculation:

$$FeO\_oxy = 0.006959 * (FeO \ salg)$$

$$MnO\_oxy = 0.000091 * ((\%MnHm * W \ Hm) - (\%Mnsteel * W \ Steel))$$

$$P\_oxy = 0.000403 * ((\%P \ Hm * W \ Hm) - (\%P \ Steel * W \ Steel))$$

$$Hm \ Si\_Oxy = 0.000444 * (\%Si \ Hm * W \ Hm)$$

$$Hm \ C\_oxy = 0.000458 * ((\%C \ Hm * W \ Hm) - (\%C \ Steel * W \ Steel))$$

$$Ore\_oxy = 0.00009392 * (\%Fe \ ore * W \ ore)$$

Total Oxygen Required $= (FeO\_Oxy + MnO\_oxy + P\_Oxy + Hm \ Si\_Oxy$
$$+ \ Hm \ C\_oxy + Ore\_oxy) * 1000 * 22.4$$

With the help of the equations and formulae, the following model has been developed.

12.3 GUI/APP IN MATLAB

The data selected from the BOF section of the integrated steelworks are analyzed with MS-Excel to obtain the result and verify the existing model of this plant. Subsequently, software (GUI/APP) was developed. Figure 12.2 shows the GUI/APP interface. One can notice that there are some values added in front of a few parameters, while several of the parameters are left blank. This GUI needs some inputs from the user, like hot metal and scrap weight, hot metal chemistry, basicity, and desired chemistry of steel.

A user needs to click on the "Calculate" button in Figure 12.2 and it will provide values in front of other blank parameters. Figure 12.3 shows the GUI/APP interface with the user-defined inputs and the model predictions.

In Figure 12.3, an operator can get an idea regarding the additions that can be made in order to maintain the desired chemistry that was defined by the user.

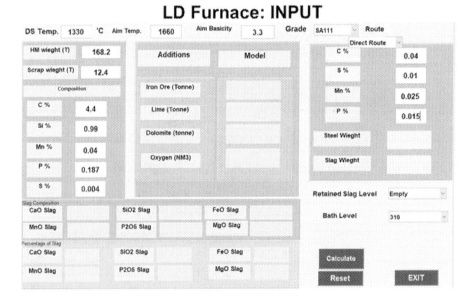

FIGURE 12.2 GUI/APP for LD furnace: Inputs needed for this model.

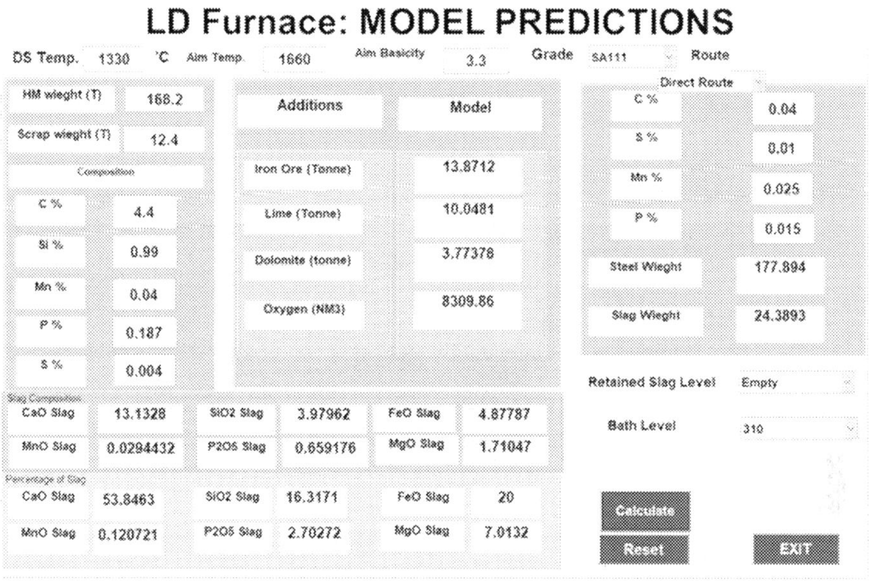

LD Furnace: MODEL PREDICTIONS

| DS Temp. | 1330 | °C | Aim Temp. | 1660 | Aim Basicity | 3.3 | Grade | SA111 | Route |

					Direct Route
HM wieght (T)	168.2		Additions	Model	C %	0.04
Scrap wieght (T)	12.4				S %	0.01
Composition			Iron Ore (Tonne)	13.8712	Mn %	0.025
C %	4.4		Lime (Tonne)	10.0481	P %	0.015
Si %	0.99		Dolomite (tonne)	3.77378	Steel Wieght	177.894
Mn %	0.04		Oxygen (NM3)	8309.86		
P %	0.187				Slag Wieght	24.3893
S %	0.004					

Slag Composition								Retained Slag Level	Empty
CaO Slag	13.1328	SiO2 Slag	3.97962	FeO Slag	4.87787			Bath Level	310
MnO Slag	0.0294432	P2O5 Slag	0.659176	MgO Slag	1.71047				

Percentage of Slag						
CaO Slag	53.8463	SiO2 Slag	16.3171	FeO Slag	20
MnO Slag	0.120721	P2O5 Slag	2.70272	MgO Slag	7.0132

Calculate
Reset
EXIT

FIGURE 12.3 GUI/APP for LD furnace: User-defined inputs along with the model predictions regarding additions and slag content.

12.4 SIMULATION OF LD STEELMAKING PROCESS IN THERMO-CALC

Process metallurgy module in Thermo-Calc 2021b was used for simulating LD steel making process under the framework of CALPHAD approach. GUI developed in this chapter does not take into account the kinetics of the steel making process. Though the additions predicted through a GUI (Figure 12.3) for the given input (Figure 12.2) are in accordance with the automation model used in the industries. Thus, we used the input (Figure 12.2) and the additions (Figure 12.3) for developing a kinetic model for LD steelmaking furnace in Thermo-Calc software.

As per our expertise and the information obtained from the industry, few assumptions were made to calibrate the model. These assumptions are listed as follows:

- Reaction time: A typical LD process cycle lasts about 35 to 40 minutes, usually referred as tap to tap time. In this work, the cycle time was fixed at 35 minutes.
- Oxygen blowing: In Figure 12.3, it has been shown that 8304 Nm^3 of oxygen is required for this cycle. Oxygen is blown for a part of the entire process and can last between 15 to 30 minutes, depending on furnace capacity and other industrial constraints. For the furnace under consideration, average oxygen blowing time was about 16 to 22 minutes, with a maximum of 26 minutes. For developing the CALPHAD based model, we fixed the blowing time at 28 minutes. Oxygen flow rate was 300 Nm^3/minute. In the kinetic module,

oxygen can be divided between the "reaction zone" and the "bulk". In this model, 200 Nm^3/minute was assigned for the "reaction zone" and 100 NM^3/minute oxygen was assigned to the bulk. Thus, blowing time was relaxed by two (2) minutes, and total amount of oxygen was relaxed by about 100 Nm^3.

- Scrap addition: In Figure 12.2 and 12.3, it is shown that 12.4 tons of scrap need to be added. In Thermo-Calc, 12.4 tons of scrap were added in the first ten minutes. That is, 1.240 tons of scrap per minute. Calculations were also performed when entire 12.4 tons of scrap were added at the begining, but the end result was same though the temperature profile will vary over the course of blowing in this case.

- Iron-ore, dolomite and lime addition: These additions were made mostly within the first ten to fifteen minutes, and additions were made at regular intervals. This model was developed as per our understanding of the LD-process, and thus the additions were divided into batches accordingly.

In Figure 12.2 and 12.3, aim temperature is mentioned on the GUI (1660 °C). Monitoring of the aim temperature is important for an operator. Figure 12.4 shows the variation of temperature over 35 minutes through the CALPHAD-based model developed in this work. Final temperature shown in Figure 12.4 is about 1585°C. In the actual data sheet, final temperature was reported at about 1610 °C. The temperature difference between the actual industry data and CALPHAD-based predictions are quite small as we are dealing with a complex process of steel making. Thus, this model can be utilized for further study.

In Figure 12.4, it can be seen that the temperature at the slag-metal interface in the reaction zone is higher when compared to other line plots. The temperature

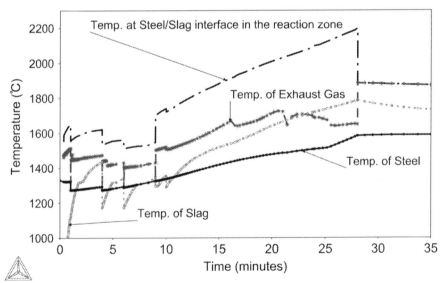

FIGURE 12.4   LPHAD-based model for LD-steel making: Variation of temperature of steel, slag, exhaust gas and the temperature at the steel slag interface over 35 minute.

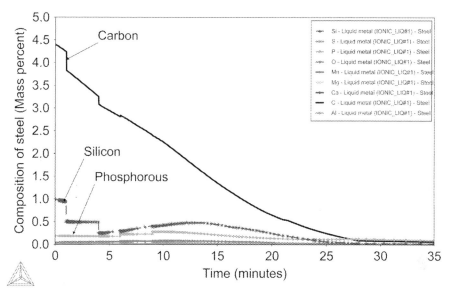

FIGURE 12.5   CALPHAD-based prediction: Variation of composition of steel bath over 35 minutes.

drops at about 28 minutes, the instant when the oxygen blowing ends. Oxidation of carbon and silicon is an exothermic process and generates heat, thus such a pattern is expected (a drop in temperature after end of blowing). Figure 12.5 shows the composition of steel over 35 minutes. This figure can also be seen as a figure depicting the removal of impurities over 35 minutes. In order to show the other impurities clearly, Figure 12.6 was plotted, which is an enlarged (zoomed) image of Figure 12.5.

An operator refers to the automation model for guidance and takes decisions based on the predictions of the automation model, their own experience with that furnace and their intuition. Monitoring of temperature of the steel bath is extremely important as it affects various downstream processes. Downstream process includes refining in the ladle and deoxidation, and it is followed by continuous casting of steel. This liquid to solid transition makes steel susceptible to cracking. Alloying elements can have a major influence on the crack susceptiblity of steel. Thus, we performed an analaysis for studying effects of various elements on crack susceptiblity coefficient in Al-Si killed microalloyed steel.

In the CALPHAD-based model, we used an an additional amount of oxygen, about 100 Nm^3 more than that predicted through the GUI. From an operators perspective, this extra oxygen will raise the oxygen content of the bath by few PPM. This additional oxygen has to be removed by adding elements like aluminum during deoxidation process. This will result in the floatation of inclusions and can affect the quality of steel. Thus, we performed simulations within the framework of CALPHAD approach for determining susceptibility of cracking in Al-Si killed steel.

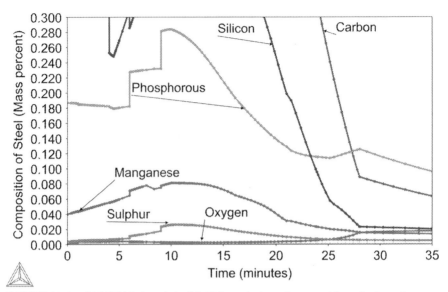

FIGURE 12.6 CALPHAD-based model: Enlarged view of composition of minor elements in the steel over 35 minutes.

12.5 CALPHAD APPROACH: DETERMINING EFFECT OF VARIOUS MICROALLOYING ELEMENTS ON CRACK SUCCEPTIBLITY COEFFICIENT OF AL-SI KILLED MICROALLOYED STEEL

During continuous casting, liquid steel solidifies and the cast product may contain various defects. The most prominent defect is due to segregation and is caused as a result of precipitation of various inclusions and precipitates. This results in formation of transverse cracks on the surface of the casting. Microalliying elements like Nb, V, B; are added for enhancing various mechanical properties of steel, but these elements are often responsible for these transverse cracks. Microalloying elements are responsible for these cracks when the temperature of the processing of these steels is between 700 and 900°C.

In this section, a class of Al-Si based killed steel was studied within the framework of CALPHAD approach. Table 12.1 shows the variable bounds for various alloying elements and the temperature considered for this study. Base steel comprises of elements like C, Si, Mn, P, S, Al and balance Fe, while Nb, Ti, V, and B are the microalloying elements. In this work, Thermo-Calc 2021b was used for determining the crack susceptiblity coefficient (CSC) of the base steel in the beginning. It was followed by estimation of CSC for base steel and Nb as the microalloying element. Thereafter, V was included along with Nb, and it was followed by Nb + V + Ti and final lot included Nb, V, Ti and B.

Resulting data were analyzed through a smoothing spline ANOVA algorithm in ESTECO modeFRONTIER software and the result have been plotted in Figure 12.7 Through statistical analysis, one can observe that titanium strongly affects the CSC.

TABLE 12.1

Variable Bounds for Composition and Temperature for Al-Si Killed Base Steel (C, Si, Mn, P, S, Al), along with Microalloying Elements Like Nb, V, Ti, B

Sl. No.	Parameters	Min.	Max.
1.	C	0.12	0.15
2.	Si	0.21	0.34
3.	Mn	1.32	1.52
4.	P	0.016	0.020
5.	S	0.028	0.034
6.	Al	0.015	0.060
7.	Nb	0.0365	0.0425
8.	V	0.05	0.064
9.	Ti	0.020	0.030
10.	B	36 PPM	40 PPM
11.	Fe	Balance	Balance
12.	Temperature	600	1000

Elements like Nb, B, and V, too affect the CSC to a certain extent. Thus, we studied the data through AI-based approach for better understanding.

All the data was scaled and analyzed through the concept of self-organizing maps (SOMs). Figure 12.8 shows the component plot obtained through SOM analysis, while Figure 12.9 shows the distribution of microalloying elements Nb, V, Ti, and B along with the temperature in the SOM space. In Figure 12.8, all the elements of the base steel are clustered together along with the temperature. CSC, Ti and B are close to each other. Nb and V are in a different region on the SOM space. In Figure 12.9, one can observe that the region of maximum titanium and

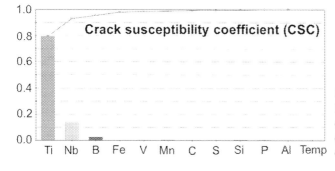

FIGURE 12.7 ANOVA: Effect of addition of microalloying element on crack susceptibility coefficient (CSC).

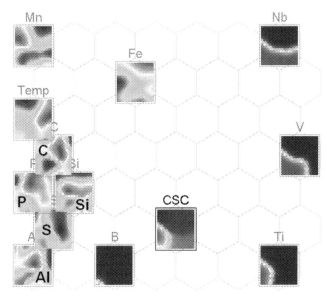

FIGURE 12.8 SOM analysis: SOM component plot showing orientation of various alloying elements, temperature and crack susceptibility coefficient of steel.

CSC coincide with each other on the SOM space. Nb, V, Ti and B are microalloying elements, and there were lots of missing values; still SOM analysis was able to determine correlations and patterns in the dataset that can be verified through the literature on crack susceptibility in steel. In Figure 12.9, temperature has also been included as most of the defects in microalloyed steel is due to processing in the temperature range of 700 to 900°C during continuous casting of steel. In Figure 12.9, one can observe that crack susceptibility of steel is highest in average temperature range in the SOM space, that is close to 700 to 900°C. CSC decreases for upper and lower temperature bounds.

Through Figure 12.8 and 12.9, one can estimate the effect of various micro-alloying elements on the crack susceptibility in steel within the framework of CALPHAD approach. Reported literature on crack susceptibility in steel during continuous casting lists elements like Nb, V, B etc as elements responsible for transverse cracks. In this study too, an increase in Ti, Nb, B, and V leads to an increase in the crack susceptibility coefficient in the temperature range between 700 and 900°C. Thus, CALPHAD-based tool along with statistical and AI-based models can be helpful in determining patterns observed during continuous casting of steel with respect to formation of transverse cracks. An operator can get vital insights for improving hot ductility of steel with respect to composition of steel and the operation temperature. An operator can design manufacturing protocol that can prove to be helpful in avoiding/minimizing the formation of transverse cracks by controlling the operating temperature and composition of steel.

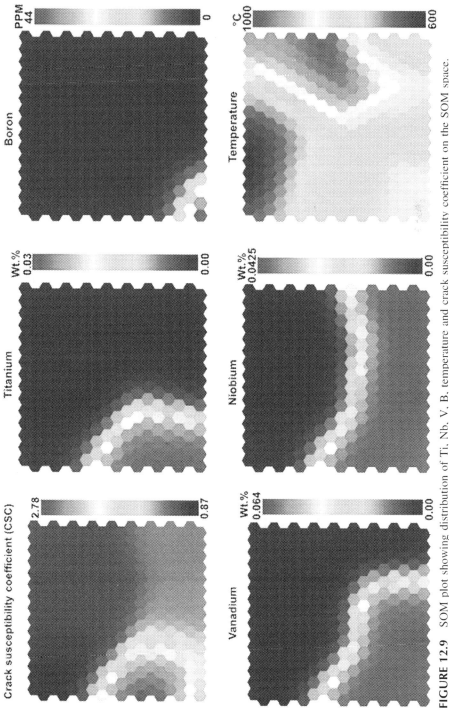

FIGURE 12.9 SOM plot showing distribution of Ti, Nb, V, B, temperature and crack susceptibility coefficient on the SOM space.

12.6 CONCLUSION

In this case study, we used actual furnace data to solve this problem analytically. We performed mass balance and did not include heat balance in this work. Our model predictions are in accordance with the furnace operations. Finally, we developed a graphical user interface (GUI) or APP in MATLAB programming language. We have presented all the governing equations in this chapter that helped us in predicting the additions to be made in the LD furnace. A user can easily follow these equations and develop the GUI/APP in MATLAB. They can even insert this equation in an Excel sheet and use a data point shown presented in this work.

- During the study of BOF process analysis on the basis of quality, cost, and productivity, we found that temperature control is the main challenge in the manufacture of BOF steel.
- Temperature monitoring is directly related to the addition we make during the blowing process. The addition of various inputs, therefore, had to be verified in order to obtain a better output.
- An automation model plays a significant role during the BOF process, which predicts the additions and oxygen to be blown during the blowing process.
- Heat and mass balance calculations are an effective method for predicting inputs.
- In GUI, only the mass balance was considered. Using mass balance, we were able to predict iron-ore, lime, dolomite addition with steel weight, slag weight, and composition. We also anticipated the amount of oxygen that would be blown during the operation. Predictions for various additions obtained through the analytical model is in accordance with the automation model in the industry.
- CALPHAD-based simulations: Kinetic simulation of the LD steel making furnace was performed within the framework of CALPHAD approach using the inputs and additions predicted through the analytical model (GUI). Temperature and composition of steel predicted through the CALPHAD based kinetic model of LD-furnace is in accordance with the industrial findings.
- The crack susceptibility coefficient for Al-Si based killed microalloyed steel: In this work, effect of the addition of various alloying elements on the crack susceptibility of Al-Si based killed steel was estimated within the framework of CALPHAD approach. Reported results are in accordance with industrial findings. Combined CALPHAD-AI based analysis can prove to be helpful in determining the optimum composition and temperature that can be helpful in improving hot ductility of steel.

13 Case Study #13: Selection of a Supervised Machine Learning (Response Surface) Algorithm for a Given Problem

13.1 INTRODUCTION: BACKGROUND

In this work, we made an attempt to determine the accuracy of the response surface predictions outside the variable space from which it was developed.

Consider a given function: $Y = X + 0.5 * Sin(5 * X)$. Figure 13.1 shows the plot for that function for two different data ranges, which have been tabulated in Table 13.1. In Figure 13.1, a user can observe that it is a normal plot. On preliminary examination, any user will not think of using supervised machine learning (ML) approaches to fit this data as it can be easily fit to a curve.

We have presented this curve as it is just a one-dimensional problem, or a problem where function depends on one variable.

Moving ahead, we applied a set of supervised machine learning (ML) approaches to fit the data for this function where X varies from 1 to 9. As usual, the fitting will be nice, but then the variable range changes to 0.1–9.9 for X. Then these models were used over the extrapolated variable range. One can observe that the model predictions are not as per our expectation, as shown in Figures 13.2 and 13.3 when we compare them with Figure 13.1. Data is given in Table 13.1 for users who want to practice this problem.

In both Figures 13.2 and 13.3, a user can observe that all of these supervised ML-based approaches perform well between 1 and 9. These models were trained for Y as a function of X, where X varied from 1 to 9, but all of these models' predictions cannot be trusted when variable bounds are relaxed. A reader can use the data in Table 13.1 and try any supervised ML algorithm; they will get similar results.

13.1.1 PURPOSE OF THIS STUDY: BENEFITS OF SUPERVISED ML-BASED PREDICTIVE MODELS

Supervised machine learning approaches are extremely helpful in developing AI/ML-based models for extremely noisy data sets. Several platforms exist that

DOI: 10.1201/9781003167372-13

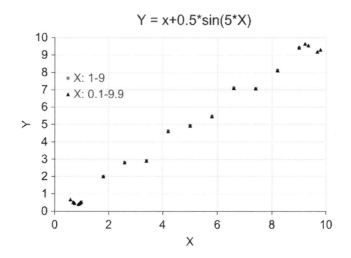

FIGURE 13.1 A problem in one dimension.

TABLE 13.1
Data Table Used for Figures 13.1, 13.2, and 13.3

X: 1–9	Y	X: 0.1–9.9	Y
1	0.520538	0.6	0.67056
1.8	2.006059	0.7	0.524608
2.6	2.810084	0.9	0.411235
3.4	2.919301	0.95	0.450354
4.2	4.618328	0.75	0.464219
5	4.933824	9	9.425452
5.8	5.468183	9.225	9.645435
6.6	7.099956	9.675	9.200333
7.4	7.078231	9.7875	9.302167
8.2	8.120689	9.3375	9.548835
9	9.425452	1	0.520538
		1.8	2.006059
		2.6	2.810084
		3.4	2.919301
		4.2	4.618328
		5	4.933824
		5.8	5.468183
		6.6	7.099956
		7.4	7.078231
		8.2	8.120689
		9	9.425452

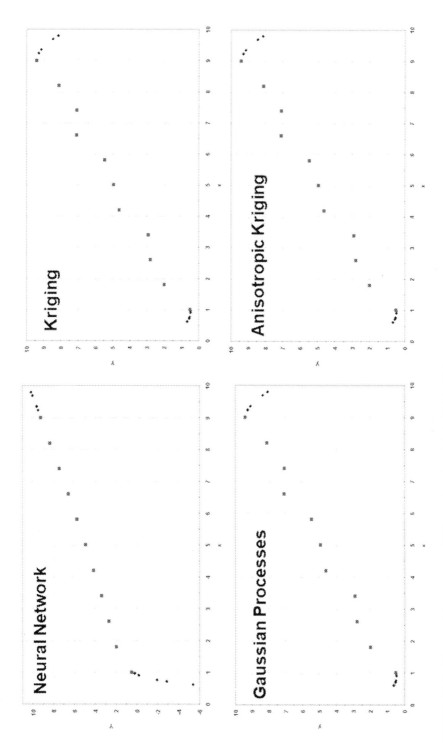

FIGURE 13.2 Fitting of few supervised ML models fromdata in Table 13.1.

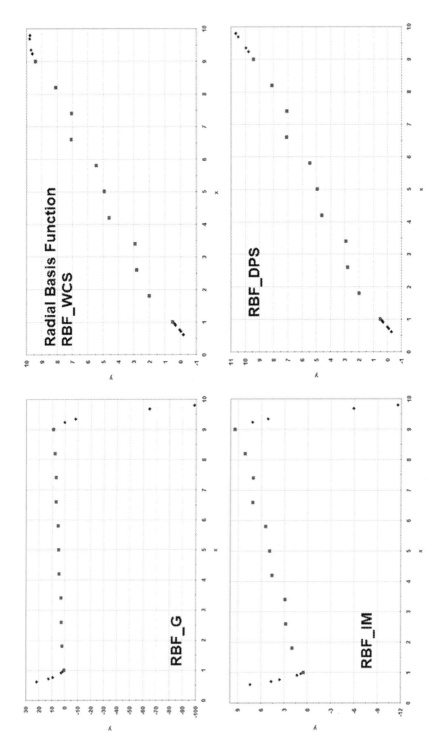

FIGURE 13.3 Fitting of supervised ML models (radial basis functions-based) from data in Table 13.1.

have been used in this book, where a user can develop extremely accurate models for extremely noisy data sets just by clicking a few buttons. Most of the computational platforms have default parameters set in such a way that a user does not even need to change any parameters. In a few minutes, they can develop an accurate model that they can use for future work as a predictive tool.

13.1.2 PURPOSE OF THIS STUDY: CHALLENGES OR LIMITATIONS

Supervised ML-based models perform well within the variable range they were trained on. On relaxing the variable bounds, all of these models suffer from some degree of uncertainty. Thus, parametrization or hyper-tuning of model parameters is the key. As mentioned previously, several computational platforms have default values that will provide an accurate model, but for that case study. In Figure 13.1, a simple 1-D equation has been fit and model predictions show irregularities. One can understand what can happen if they work with multi-dimensional problems.

In this chapter, we have presented several case studies through which a reader can understand the challenges associated with developing AI/ML-based models and quantify their limitations. At the same time, a reader will also get vital information on which models to consider for a problem with a given number of variables.

The most important part of developing predictive models through AI/ML techniques is to have an understanding of the problem and the related techniques that a user wants to use. Some problems can be sorted easily by curve fitting; thus, using AI/ML techniques will be a waste of time and resources for those problems. One can try any algorithm, just like we have done in this chapter. In this chapter, we have provided a set of training data for two-dimensional problems in Table 13.2. Additionally, we have provided error metrics values and the functions for 24 Schittkowski test cases that a user can work on just by reading this chapter.

13.2 METHOD

For this work, the response surfaces were developed by different modules available in the commercial optimization package modeFRONTIER. The test cases used in this study are the Schittkowski test cases. We first tested the 2-D cases (cases 1 through 24).

Since we are dealing with supervised machine learning approaches, we need training data for training the model and testing data for testing the model predictions. For generating a training data set, a scarce data set (of 20 points) in the range of (0–9) was generated by Sobol's algorithm. This was used to train (develop) the response surface for the 24 test cases. For generating a testing data set, the bounds were relaxed by 10% around the existing variable space (that is, $0.1 * (x0,x1)_{min} - 1.1 * (x0,x1)_{max}$). In this case, it will be from 0.9 to 9.9. Sobol's algorithm was again used to generate data in the extended bounds. Table 13.2 provides the training and testing data set that is to be used for the 24 test problems. Distribution of data in the variable space can be observed through Figures 13.4 (training data) and 13.5 (testing data).

TABLE 13.2

Training and Testing Data Set for Developing and Testing the Response Surfaces for the 24 Schittkowski Test Cases

Training Set (Bounded)		Testing Set (Extended Bound)	
x0	x1	x0	x1
1	1	0.9	0.9
3	7	0.925	7.65
7	3	0.975	3.15
8	8	0.9875	8.775
4	4	0.9375	4.275
2	6	9	0.9
6	2	9.225	7.65
6.5	7.5	9.675	3.15
2.5	3.5	9.7875	8.775
4.5	5.5	9.3375	4.275
8.5	1.5	3.15	0.975
7.5	6.5	7.65	0.925
3.5	2.5	8.775	0.9875
1.5	8.5	4.275	0.9375
5.5	4.5	2.025	0.9625
5.75	8.75	0.9	9
1.75	4.75	3.15	9.675
3.75	6.75	7.65	9.225
7.75	2.75	8.775	9.7875
8.75	5.75	4.275	9.3375

13.2.1 MODULES USED IN MODEFRONTIER

It must be noted that only DEFAULT model parameters were used for these studies. This was done to demonstrate the importance of parametrization or tuning of parameters.

1. **AKR:** Cartesian anisotropic kriging regression using
   a. Gaussian (**AKR_G**)
   b. Exponential (**AKR_Exp**)
2. **GP:** Response surface based on the Gaussian processes algorithm
   a. Maximize the likelihood function, avoids overfitting (**GP_maxlik**)
   b. Minimize the interpolation errors, more robust solution (**GP_miner**)
3. **KN:** Response surface based on Sheppard method and its generalizations (**KN**)
4. **Kriging**
   a. Gaussian (**KR_G**)
   b. Exponential (**KR_Exp**)

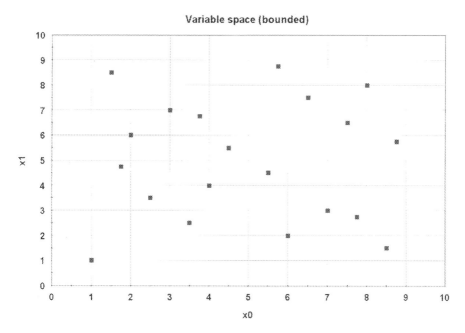

FIGURE 13.4 Training set: Distribution of variables in variable space for data in Table 13.2.

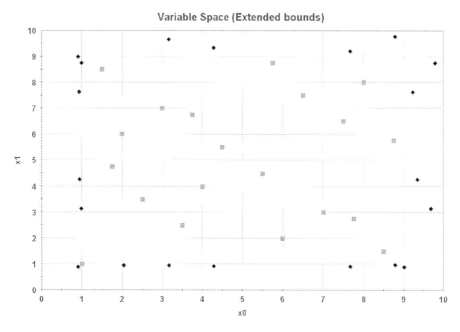

FIGURE 13.5 Testing set with extended bounds: Distribution of variables in variable space for data in Table 13.2.

5. **RBF:** Radial basis function:
   a. **HM:** Hardy's multiquadrics (**RBF_HM**)
   b. **IM:** Inverse multiquadrics (**RBF_IM**)
   c. **Gauss:** Gaussian (**RBF_G**)
   d. **DPS:** Duchon's polyharmonic splines (**RBF_DPS**)
   e. **WCS:** Wedland's compactly supported (**RBF_WCS**)

13.2.2 Accuracy Measurements

Accuracy measurements of the trained (generated) RSM were performed over the testing data set tabulated in Table 13.2. Three different error metrics were used: R Square, Relative Average Absolute Error (RAAE), and Relative Maximum Absolute Error (RMAE). The calculated accuracy measures are tabulated in the following text. Tables 13.3–13.26 will provide a better understanding of the error metrics associated with each of these models.

13.3 RESULTS: ERROR METRICS FOR VARIOUS SUPERVISED MACHINE LEARNING MODELS

In this section, performance of several response surfaces has been evaluated over data that were within the variable bounds, as well as over data that is outside the variable bounds. Thereafter, we have provided with a list of suggested response surface that a user can use if they are working with similar data or a test problem.

TABLE 13.3
Error Metrics over In-bound and Relaxed Variable Bounds for Schittkowski Test Case 1

Case 1: pow(x[1] – x[0] * x[0], 2.) * 100. + pow(1. – x[0], 2.)

	R-Square Bound	R-Square Extended	RAAE Bound	RAAE Extended	RMAE Bound	RMAE Extended
AKR_G	1	1.0000	3.91E-08	0.0016	9.38E-08	0.0083
AKR_Exp	1	0.8073	2.09E-15	0.3914	6.59E-15	1.7161
KR_G	1	1.0000	3.03E-08	0.0016	7.51E-08	0.0083
KR_Exp	1	0.3389	1.45E-15	1.6001	9.16E-15	5.3298
GP_maxlik	1	0.9870	7.20E-09	0.0706	2.52E-08	0.3734
GP_miner	1	0.8726	2.92E-14	0.2679	1.24E-13	1.3205
KN	1	0.4539	0.00E+00	1.0305	0.00E+00	3.8602
RBF_HM	1	0.9947	5.64E-13	0.0432	1.58E-12	0.2462
RBF_IM	1	0.9823	9.23E-14	0.0802	3.13E-13	0.4659
RBF_G	1	0.9965	1.93E-12	0.0386	4.41E-12	0.1925
RBF_DPS	1	0.9489	3.44E-16	0.1646	3.73E-15	0.7477
RBF_WCS	1	0.7077	7.65E-16	0.4528	4.86E-15	2.3126

TABLE 13.4

Error Metrics over In-bound and Relaxed Variable Bounds for Schittkowski Test Case 2

Case 2: pow(x[1] − x[0] * x[0], 2.) * 100. + pow(1. − x[0], 2.) − (0.0504261879)

	R-Square Bound	R-Square Extended	RAAE Bound	RAAE Extended	RMAE Bound	RMAE Extended
AKR_G	1	1.0000	4.00E-08	0.0016	1.09E-07	0.0083
AKR_Exp	1	0.5192	3.14E-16	0.9634	8.10E-16	3.6589
KR_G	1	1.0000	3.48E-08	0.0016	7.04E-08	0.0083
KR_Exp	1	0.8073	2.44E-15	0.3914	6.10E-15	1.7161
GP_maxlik	1	0.9519	4.11E-10	0.1526	1.38E-09	0.7575
GP_miner	1	0.8110	1.86E-14	0.3962	4.37E-14	1.5300
KN	1	0.5421	2.10E-16	0.7841	7.29E-16	3.1513
RBF_HM	1	0.9947	5.64E-13	0.0432	1.58E-12	0.2462
RBF_IM	1	0.9823	9.23E-14	0.0802	3.13E-13	0.4659
RBF_G	1	0.9965	1.93E-12	0.0386	4.41E-12	0.1925
RBF_DPS	1	0.9489	4.70E-16	0.1646	4.17E-15	0.7477
RBF_WCS	1	0.7077	6.82E-16	0.4528	5.08E-15	2.3126

TABLE 13.5

Error Metrics over In-bound and Relaxed Variable Bounds for Schittkowski Test Case 3

Case 3: x[1] + pow(x[1] − x[0], 2.) * 1e-5

	R-Square Bound	R-Square Extended	RAAE Bound	RAAE Extended	RMAE Bound	RMAE Extended
AKR_G	1	1.0000	8.14E-10	0.0000	1.88E-09	0.0001
AKR_Exp	1	0.9743	8.04E-16	0.1430	3.32E-15	0.4471
KR_G	1	1.0000	8.14E-10	0.0000	1.88E-09	0.0001
KR_Exp	0.9999981	0.9223	7.00E-04	0.2911	2.66E-03	0.8340
GP_maxlik	1	1.0000	4.59E-11	0.0006	1.04E-10	0.0020
GP_miner	1	0.9806	5.19E-15	0.1205	1.82E-14	0.3295
KN	1	0.8644	0.00E+00	0.4573	0.00E+00	1.0189
RBF_HM	1	1.0000	4.27E-14	0.0041	1.03E-13	0.0172
RBF_IM	1	0.9990	2.99E-14	0.0211	9.61E-14	0.0967
RBF_G	1	0.9999	1.85E-13	0.0091	5.88E-13	0.0329
RBF_DPS	1	1.0000	4.09E-16	0.0000	2.74E-15	0.0000
RBF_WCS	1	0.8272	1.41E-15	0.3583	3.62E-15	1.4788

TABLE 13.6

Error Metrics over In-bound and Relaxed Variable Bounds for Schittkowski Test Case 4

Case 4: pow(x[0] + 1., 3.) / 3. + x[1] – (8/3)

	R-Square Bound	R-Square Extended	RAAE Bound	RAAE Extended	RMAE Bound	RMAE Extended
AKR_G	1	1.0000	5.93E-08	0.0002	1.80E-07	0.0010
AKR_Exp	1	0.8975	3.83E-15	0.2675	1.05E-14	1.1533
KR_G	1	1.0000	8.33E-08	0.0002	1.81E-07	0.0010
KR_Exp	1	0.8975	2.12E-15	0.2675	7.82E-15	1.1533
GP_maxlik	1	0.9429	1.02E-07	0.1872	2.27E-07	0.8217
GP_miner	1	0.8766	7.40E-15	0.2876	3.14E-14	1.2944
KN	1	0.7072	1.59E-15	0.6000	6.94E-15	2.2310
RBF_HM	1	0.9989	1.39E-13	0.0188	2.87E-13	0.1160
RBF_IM	1	0.9935	4.09E-14	0.0456	1.27E-13	0.2793
RBF_G	1	0.9990	5.35E-13	0.0198	1.49E-12	0.0997
RBF_DPS	1	0.9848	1.72E-15	0.0962	6.47E-15	0.3772
RBF_WCS	1	0.7549	2.59E-15	0.3980	8.78E-15	2.0119

TABLE 13.7

Error Metrics over In-bound and Relaxed Variable Bounds for Schittkowski Test Case 5

Case 5: sin(x[0] + x[1]) + pow(x[0] – x[1], 2.) – x[0] * 1.5 + x[1] * 2.5 + 1. – (–0.5 * sqrt(3)-pi/3.)

	R-Square Bound	R-Square Extended	RAAE Bound	RAAE Extended	RMAE Bound	RMAE Extended
AKR_G	0.9999987	0.9705	3.81E-04	0.1046	1.63E-03	0.5079
AKR_Exp	1	0.8875	3.70E-15	0.3423	1.23E-14	1.0410
KR_G	1	0.9714	1.01E-13	0.1126	2.27E-13	0.4884
KR_Exp	1	0.8875	4.39E-15	0.3423	1.09E-14	1.0410
GP_maxlik	1	0.9772	1.61E-08	0.1162	4.34E-08	0.3048
GP_miner	1	0.8014	5.59E-07	0.4910	2.56E-06	1.2584
KN	1	0.7044	2.13E-15	0.6330	4.23E-15	1.7920
RBF_HM	1	0.9836	8.69E-13	0.0783	2.15E-12	0.3531
RBF_IM	1	0.9892	7.47E-14	0.0776	2.29E-13	0.2584
RBF_G	1	0.9338	2.69E-12	0.1307	9.00E-12	0.7071
RBF_DPS	1	0.9789	1.54E-15	0.1225	2.91E-15	0.3558
RBF_WCS	1	0.8868	1.64E-15	0.3011	3.17E-15	0.9423

TABLE 13.8

Error Metrics over In-bound and Relaxed Variable Bounds for Schittkowski Test Case 6

Case 6: pow(1. – x[0], 2.)

	R-Square Bound	R-Square Extended	RAAE Bound	RAAE Extended	RMAE Bound	RMAE Extended
AKR_G	1	0.8493	1.87E-15	0.3688	4.48E-15	1.4489
AKR_Exp	1	1.0000	3.66E-09	0.0001	8.26E-09	0.0003
KR_G	1	0.9839	4.79E-09	0.0989	1.01E-08	0.3990
KR_Exp	1	0.9081	7.21E-15	0.2479	3.19E-14	1.0411
GP_maxlik	1	0.7654	0.00E+00	0.5190	0.00E+00	1.8178
GP_miner	1	0.8493	6.20E-13	0.3688	2.29E-12	1.4487
KN	1	1.0000	3.26E-09	0.0001	6.84E-09	0.0003
RBF_HM	1	0.9913	5.35E-16	0.0804	3.30E-15	0.2401
RBF_IM	1	0.9995	1.36E-13	0.0134	5.77E-13	0.0660
RBF_G	1	0.9995	8.92E-14	0.0145	3.05E-13	0.0704
RBF_DPS	1	0.9962	2.22E-14	0.0364	6.56E-14	0.2056
RBF_WCS	1	0.7869	1.37E-15	0.3594	4.32E-15	1.7748

TABLE 13.9

Error Metrics over In-bound and Relaxed Variable Bounds for Schittkowski Test Case 7

Case 7: log(pow(x[0], 2.) + 1.) – x[1] – (-sqrt(3))

	R-Square Bound	R-Square Extended	RAAE Bound	RAAE Extended	RMAE Bound	RMAE Extended
AKR_G	1	0.9735	1.31E-15	0.1667	2.79E-15	0.3297
AKR_Exp	1	0.9735	1.31E-15	0.1667	2.79E-15	0.3297
KR_G	1	0.9996	1.20E-08	0.0151	4.29E-08	0.0656
KR_Exp	1	0.9569	1.17E-14	0.1923	4.05E-14	0.4736
GP_maxlik	1	0.9735	1.31E-15	0.1667	2.79E-15	0.3297
GP_miner	0.999907	0.9271	4.95E-03	0.3196	2.39E-02	0.5982
KN	1	1.0000	3.13E-11	0.0052	9.34E-11	0.0172
RBF_HM	1	0.9735	1.31E-15	0.1667	2.79E-15	0.3297
RBF_IM	1	0.9998	4.86E-13	0.0112	1.17E-12	0.0282
RBF_G	1	0.9735	1.31E-15	0.1667	2.79E-15	0.3297
RBF_DPS	1	0.9993	7.62E-14	0.0224	1.91E-13	0.0508
RBF_WCS	1	0.9996	1.20E-08	0.0151	4.29E-08	0.0656

TABLE 13.10

Error Metrics over In-bound and Relaxed Variable Bounds for Schittkowski Test Case 9

Case 9: sin(pi * x[0] / 12.) * cos(pi * x[1] / 16.) – (–0.5)

	R-Square Bound	R-Square Extended	RAAE Bound	RAAE Extended	RMAE Bound	RMAE Extended
AKR_G	1	0.9392364	6.81E-15	0.2371189	2.16E-14	0.5499582
AKR_Exp	1	0.9999946	1.49E-11	0.0013325	4.01E-11	0.007784
KR_G	0.9999999	0.9843788	0.0002225	0.0988881	0.000543	0.2610092
KR_Exp	1	0.9428015	1.14E-13	0.2124166	2.71E-13	0.6292054
GP_maxlik	1	0.763916	8.07E-15	0.5735795	2.70E-14	1.4040153
GP_miner	1	0.4366992	1.55E-13	1.6413992	3.89E-13	3.8256104
KN	1	0.9999915	8.74E-12	0.0016645	2.35E-11	0.0108063
RBF_HM	1	0.9837569	5.70E-15	0.1086586	1.87E-14	0.2771305
RBF_IM	1	0.9998381	1.79E-13	0.0093568	5.54E-13	0.0283645
RBF_G	1	0.9986876	3.77E-13	0.0266759	9.98E-13	0.0992585
RBF_DPS	1	0.9976246	4.38E-14	0.0392096	1.35E-13	0.111296
RBF_WCS	1	0.9407256	6.84E-15	0.2156776	2.29E-14	0.7703747

TABLE 13.11

Error Metrics over In-bound and Relaxed Variable Bounds for Schittkowski Test Case 10

Case 10: x[0] – x[1] – (–1.)

	R-Square Bound	R-Square Extended	RAAE Bound	RAAE Extended	RMAE Bound	RMAE Extended
AKR_G	1	0.9456	8.13E-14	0.2587	4.27E-13	0.5076
AKR_Exp	1	1.0000	1.35E-09	0.0000	3.26E-09	0.0001
KR_G	1	1.0000	2.85E-09	0.0040	9.94E-09	0.0099
KR_Exp	1	0.9418	1.37E-14	0.2464	4.32E-14	0.6029
GP_maxlik	1	0.9033	0.00E+00	0.3818	0.00E+00	0.7392
GP_miner	1	0.9831	1.57E-15	0.1272	4.42E-15	0.2609
KN	1	1.0000	1.63E-09	0.0000	5.04E-09	0.0001
RBF_HM	1	1.0000	9.29E-16	0.0000	1.95E-15	0.0000
RBF_IM	1	0.9999	2.44E-13	0.0080	5.60E-13	0.0243
RBF_G	1	1.0000	1.02E-13	0.0026	2.11E-13	0.0055
RBF_DPS	1	0.9996	4.23E-14	0.0154	1.03E-13	0.0416
RBF_WCS	1	0.8861	4.59E-16	0.2945	2.35E-15	0.8044

TABLE 13.12

Error Metrics over In-bound and Relaxed Variable Bounds for Schittkowski Test Case 11

Case 11: pow(x[0] − 5., 2.) + pow(x[1], 2.) − 25. − (−8.498464223)

	R-Square Bound	R-Square Extended	RAAE Bound	RAAE Extended	RMAE Bound	RMAE Extended
AKR_G	1	0.9088	2.25E-15	0.2511	6.36E-15	0.9996
AKR_Exp	1	1.0000	1.95E-09	0.0001	5.94E-09	0.0006
KR_G	1	0.9746	4.64E-09	0.1251	9.23E-09	0.5579
KR_Exp	1	0.8928	1.25E-14	0.2693	2.59E-14	1.0500
GP_maxlik	1	0.7059	4.51E-17	0.6376	3.93E-16	2.1123
GP_miner	1	0.9088	1.98E-15	0.2511	4.19E-15	0.9996
KN	1	1.0000	3.55E-09	0.0001	7.64E-09	0.0006
RBF_HM	1	0.9906	4.95E-16	0.0649	2.65E-15	0.2730
RBF_IM	1	0.9992	4.12E-13	0.0171	8.66E-13	0.0988
RBF_G	1	0.9993	8.35E-14	0.0147	2.19E-13	0.0842
RBF_DPS	1	0.9948	3.72E-14	0.0416	9.80E-14	0.2345
RBF_WCS	1	0.7272	1.73E-15	0.4692	3.56E-15	2.0001

TABLE 13.13

Error Metrics over In-bound and Relaxed Variable Bounds for Schittkowski Test Case 12

Case 12: x[0] * x[0] * 0.5 + x[1] * x[1] − x[0] * x[1] − x[0] * 7. − x[1] * 7. − (−30.)

	R-Square Bound	R-Square Extended	RAAE Bound	RAAE Extended	RMAE Bound	RMAE Extended
AKR_G	1	0.9459	1.82E-15	0.2258	8.52E-15	0.5795
AKR_Exp	1	1.0000	5.06E-09	0.0001	1.07E-08	0.0004
KR_G	1	0.9962	2.53E-08	0.0478	7.62E-08	0.1530
KR_Exp	1	0.8942	1.01E-14	0.3488	3.13E-14	0.9362
GP_maxlik	1	0.8120	0.00E+00	0.5782	0.00E+00	1.0691
GP_miner	1	0.7719	1.84E-08	0.7093	7.82E-08	1.6041
KN	1	1.0000	5.04E-09	0.0001	1.31E-08	0.0004
RBF_HM	1	0.9935	1.16E-15	0.0615	3.46E-15	0.1739
RBF_IM	1	0.9995	2.86E-13	0.0193	8.29E-13	0.0419
RBF_G	1	0.9997	1.39E-13	0.0147	3.35E-13	0.0459
RBF_DPS	1	0.9984	5.20E-14	0.0317	1.55E-13	0.1077
RBF_WCS	1	0.9171	2.27E-16	0.2694	3.97E-15	0.7385

TABLE 13.14

Error Metrics over In-bound and Relaxed Variable Bounds for Schittkowski Test Case 13

Case 13: pow(x[0] − 2., 2.) + x[1] * x[1] − (1.)

	R-Square Bound	R-Square Extended	RAAE Bound	RAAE Extended	RMAE Bound	RMAE Extended
AKR_G	1	0.9038	1.60E-15	0.2535	5.57E-15	1.0639
AKR_Exp	1	1.0000	4.37E-10	0.0001	8.16E-10	0.0005
KR_G	1	0.9720	1.74E-07	0.1229	4.22E-07	0.4895
KR_Exp	1	0.8882	1.16E-14	0.2849	2.84E-14	1.1593
GP_maxlik	1	0.6960	0.00E+00	0.6916	0.00E+00	2.3288
GP_miner	1	0.9038	1.06E-15	0.2535	2.79E-15	1.0639
KN	1	1.0000	2.71E-10	0.0001	1.01E-09	0.0005
RBF_HM	1	0.9930	5.84E-16	0.0560	2.29E-15	0.2354
RBF_IM	1	0.9992	1.65E-13	0.0174	5.44E-13	0.0946
RBF_G	1	0.9993	5.67E-14	0.0142	1.84E-13	0.0805
RBF_DPS	1	0.9942	2.63E-14	0.0436	6.66E-14	0.2434
RBF_WCS	1	0.6595	1.33E-15	0.5345	3.29E-15	2.4000

TABLE 13.15

Error Metrics over In-bound and Relaxed Variable Bounds for Schittkowski Test Case 14

Case 14: pow(x[0] − 2., 2.) + pow(x[1] − 1., 2.) − (9.-2.875 * sqrt(7))

	R-Square Bound	R-Square Extended	RAAE Bound	RAAE Extended	RMAE Bound	RMAE Extended
AKR_G	1	0.4867	2.06E-05	1.1664	6.90E-05	3.9837
AKR_Exp	1	1.0000	3.98E-10	0.0001	1.24E-09	0.0006
KR_G	1	0.9981	3.32E-07	0.0294	1.28E-06	0.1225
KR_Exp	1	0.8684	1.07E-14	0.3155	2.69E-14	1.3038
GP_maxlik	1	0.6606	1.58E-15	0.7373	4.46E-15	2.5183
GP_miner	1	0.6800	2.72E-07	0.6597	9.30E-07	2.5104
KN	1	1.0000	4.00E-10	0.0001	9.18E-10	0.0006
RBF_HM	1	0.9907	1.01E-15	0.0653	2.67E-15	0.2744
RBF_IM	1	0.9990	1.73E-13	0.0190	5.11E-13	0.1031
RBF_G	1	0.9992	1.15E-13	0.0161	3.20E-13	0.0898
RBF_DPS	1	0.9932	2.48E-14	0.0477	6.37E-14	0.2635
RBF_WCS	1	0.6420	1.38E-15	0.5574	3.86E-15	2.4924

TABLE 13.16

Error Metrics over In-bound and Relaxed Variable Bounds for Schittkowski Test Case 15

Case 15: 100. * pow(x[1] – x[0] * x[0], 2.) + pow(1. – x[0], 2.) – (306.5)

	R-Square Bound	R-Square Extended	RAAE Bound	RAAE Extended	RMAE Bound	RMAE Extended
AKR_G	1	0.3389	2.00E-15	1.6001	9.30E-15	5.3298
AKR_Exp	1	1.0000	3.91E-08	0.0016	9.38E-08	0.0083
KR_G	1	0.8456	3.75E-10	0.3408	1.20E-09	1.3680
KR_Exp	1	0.7806	5.99E-15	0.4146	2.33E-14	1.8354
GP_maxlik	1	0.5421	0.00E+00	0.7841	0.00E+00	3.1513
GP_miner	1	0.8073	1.46E-15	0.3914	6.59E-15	1.7161
KN	1	1.0000	3.03E-08	0.0016	7.51E-08	0.0083
RBF_HM	1	0.9489	3.44E-16	0.1646	3.73E-15	0.7477
RBF_IM	1	0.9965	1.93E-12	0.0386	4.41E-12	0.1925
RBF_G	1	0.9947	5.64E-13	0.0432	1.58E-12	0.2462
RBF_DPS	1	0.9823	9.23E-14	0.0802	3.13E-13	0.4659
RBF_WCS	1	0.7077	7.65E-16	0.4528	4.86E-15	2.3126

TABLE 13.17

Error Metrics over In-bound and Relaxed Variable Bounds for Schittkowski Test Case 16

Case 16: 100. * pow(x[1] – x[0] * x[0], 2.) + pow(1. – x[0], 2.) – (.25)

	R-Square Bound	R-Square Extended	RAAE Bound	RAAE Extended	RMAE Bound	RMAE Extended
AKR_G	1	0.3389	2.00E-15	1.6001	9.30E-15	5.3298
AKR_Exp	1	1.0000	3.91E-08	0.0016	9.38E-08	0.0083
KR_G	1	0.8456	3.75E-10	0.3408	1.20E-09	1.3680
KR_Exp	1	0.7806	5.99E-15	0.4146	2.33E-14	1.8354
GP_maxlik	1	0.5421	0.00E+00	0.7841	0.00E+00	3.1513
GP_miner	1	0.3389	1.45E-15	1.6001	9.16E-15	5.3298
KN	1	1.0000	3.03E-08	0.0016	7.51E-08	0.0083
RBF_HM	1	0.9489	3.44E-16	0.1646	3.73E-15	0.7477
RBF_IM	1	0.9965	1.93E-12	0.0386	4.41E-12	0.1925
RBF_G	1	0.9947	5.64E-13	0.0432	1.58E-12	0.2462
RBF_DPS	1	0.9823	9.23E-14	0.0802	3.13E-13	0.4659
RBF_WCS	1	0.7077	7.65E-16	0.4528	4.86E-15	2.3126

TABLE 13.18

Error Metrics over In-bound and Relaxed Variable Bounds for Schittkowski Test Case 17

Case 17: 100. * pow(x[1] − x[0] * x[0], 2.) + pow(1. − x[0], 2.) − (1.)

	R-Square Bound	R-Square Extended	RAAE Bound	RAAE Extended	RMAE Bound	RMAE Extended
AKR_G	1	0.807261	2.09E-15	0.3913736	6.59E-15	1.7161332
AKR_Exp	1	0.9999938	3.91E-08	0.0015529	9.38E-08	0.0082634
KR_G	1	0.8456259	3.75E-10	0.3407915	1.20E-09	1.3679758
KR_Exp	1	0.7806147	5.99E-15	0.414582	2.33E-14	1.835358
GP_maxlik	1	0.5421161	0	0.7840676	0	3.1513485
GP_miner	1	0.3389456	1.45E-15	1.6000752	9.16E-15	5.3297661
KN	1	0.9999938	3.03E-08	0.0015536	7.51E-08	0.0082641
RBF_HM	1	0.9488949	3.44E-16	0.1645745	3.73E-15	0.7476643
RBF_IM	1	0.9964736	1.93E-12	0.0386307	4.41E-12	0.1924868
RBF_G	1	0.9946945	5.64E-13	0.0432378	1.58E-12	0.2461951
RBF_DPS	1	0.9823048	9.23E-14	0.0801635	3.13E-13	0.4658576
RBF_WCS	1	0.7077223	7.65E-16	0.4528223	4.86E-15	2.3126255

TABLE 13.19

Error Metrics over In-bound and Relaxed Variable Bounds for Schittkowski Test Case 18

Case 18: 0.01 * x[0] * x[0] + x[1] * x[1] − (5.)

	R-Square Bound	R-Square Extended	RAAE Bound	RAAE Extended	RMAE Bound	RMAE Extended
AKR_G	1	0.9391	1.13E-15	0.1913	3.33E-15	0.8158
AKR_Exp	1	1.0000	1.35E-09	0.0001	3.12E-09	0.0004
KR_G	1	0.9965	8.32E-09	0.0420	3.04E-08	0.1810
KR_Exp	1	0.9302	4.36E-15	0.2078	1.24E-14	0.7904
GP_maxlik	1	0.7826	2.23E-16	0.5341	4.00E-15	1.7222
GP_miner	1	0.9391	1.44E-15	0.1913	4.67E-15	0.8158
KN	1	1.0000	1.67E-09	0.0001	4.15E-09	0.0004
RBF_HM	1	0.9946	4.91E-16	0.0611	2.63E-15	0.1912
RBF_IM	1	0.9996	2.14E-13	0.0117	5.86E-13	0.0696
RBF_G	1	0.9996	5.98E-14	0.0125	1.27E-13	0.0640
RBF_DPS	1	0.9971	4.49E-14	0.0313	1.08E-13	0.1883
RBF_WCS	1	0.7964	1.61E-15	0.3446	3.72E-15	1.7388

TABLE 13.20

Error Metrics over In-bound and Relaxed Variable Bounds for Schittkowski Test Case 19

Case 19: pow(x[0] − 10., 3.) + pow(x[1] − 20., 3.) − (−6961.8138755801392)

	R-Square Bound	R-Square Extended	RAAE Bound	RAAE Extended	RMAE Bound	RMAE Extended
AKR_G	1	0.6017	1.31E-16	1.2259	8.05E-16	2.2274
AKR_Exp	1	1.0000	2.12E-08	0.0002	5.36E-08	0.0005
KR_G	1	0.9996	1.56E-08	0.0168	6.16E-08	0.0472
KR_Exp	1	0.9889	7.02E-15	0.0837	2.54E-14	0.2634
GP_maxlik	1	0.8723	6.06E-17	0.4090	5.11E-16	0.9553
GP_miner	1	0.9814	3.00E-15	0.1244	4.45E-15	0.2702
KN	1	1.0000	1.54E-08	0.0002	4.76E-08	0.0005
RBF_HM	1	0.9980	7.93E-16	0.0352	3.88E-15	0.1229
RBF_IM	1	0.9998	5.88E-14	0.0100	1.78E-13	0.0279
RBF_G	1	0.9999	5.13E-14	0.0069	1.02E-13	0.0202
RBF_DPS	1	0.9992	2.13E-14	0.0215	5.21E-14	0.0631
RBF_WCS	1	0.8094	2.15E-15	0.3910	5.22E-15	1.5056

TABLE 13.21

Error Metrics over In-bound and Relaxed Variable Bounds for Schittkowski Test Case 20

Case 20: 100. * pow(x[1] − x[0] * x[0], 2.) + pow(1. − x[0], 2.) − (81.5-25. * sqrt(3))

	R-Square Bound	R-Square Extended	RAAE Bound	RAAE Extended	RMAE Bound	RMAE Extended
AKR_G	1	0.8073	2.48E-15	0.3914	5.76E-15	1.7161
AKR_Exp	1	1.0000	3.87E-08	0.0016	1.20E-07	0.0083
KR_G	1	0.8456	3.75E-10	0.3408	1.20E-09	1.3680
KR_Exp	1	0.7806	6.08E-15	0.4146	2.44E-14	1.8354
GP_maxlik	1	0.5421	6.02E-16	0.7841	1.46E-15	3.1513
GP_miner	1	0.2766	1.09E-15	1.9077	2.41E-15	6.1729
KN	1	1.0000	3.93E-08	0.0016	1.20E-07	0.0083
RBF_HM	1	0.9489	4.82E-16	0.1646	8.77E-16	0.7477
RBF_IM	1	0.9965	1.93E-12	0.0386	4.41E-12	0.1925
RBF_G	1	0.9947	5.64E-13	0.0432	1.58E-12	0.2462
RBF_DPS	1	0.9823	9.23E-14	0.0802	3.13E-13	0.4659
RBF_WCS	1	0.7077	4.81E-16	0.4528	1.13E-15	2.3126

TABLE 13.22

Error Metrics over In-bound and Relaxed Variable Bounds for Schittkowski Test Case 21

Case 21: 0.01 * x[0] * x[0] + x[1] * x[1] – 100. – (–99.96)

	R-Square Bound	R-Square Extended	RAAE Bound	RAAE Extended	RMAE Bound	RMAE Extended
AKR_G	1	0.8150	2.58E-07	0.4665	1.01E-06	1.6529
AKR_Exp	1	1.0000	1.83E-09	0.0001	4.01E-09	0.0004
KR_G	1	0.9965	8.32E-09	0.0420	3.04E-08	0.1810
KR_Exp	1	0.9302	4.39E-15	0.2078	1.24E-14	0.7904
GP_maxlik	1	0.7826	2.29E-15	0.5341	3.93E-15	1.7222
GP_miner	1	0.8131	6.09E-09	0.4711	2.39E-08	1.6647
KN	1	1.0000	1.84E-09	0.0001	4.35E-09	0.0004
RBF_HM	1	0.9946	1.04E-15	0.0611	2.53E-15	0.1912
RBF_IM	1	0.9996	2.84E-13	0.0117	7.97E-13	0.0696
RBF_G	1	0.9996	1.01E-13	0.0125	2.46E-13	0.0640
RBF_DPS	1	0.9971	3.80E-14	0.0313	9.51E-14	0.1883
RBF_WCS	1	0.7964	1.71E-15	0.3446	3.48E-15	1.7388

TABLE 13.23

Error Metrics over In-bound and Relaxed Variable Bounds for Schittkowski Test Case 22

Case 22: pow(x[0] – 2., 2.) + pow(x[1] – 1., 2.) – (1.)

	R-Square Bound	R-Square Extended	RAAE Bound	RAAE Extended	RMAE Bound	RMAE Extended
AKR_G	0.5303681	0.4107	4.03E-01	0.6922	1.20E+00	1.8395
AKR_Exp	0.5303681	0.5069	3.32E-01	0.5526	9.91E-01	1.3286
KR_G	0.5303681	0.4759	3.41E-01	0.5878	1.02E+00	1.3601
KR_Exp	0.5303681	0.4037	4.02E-01	0.7081	1.20E+00	1.8310
GP_maxlik	0.5303681	0.2762	5.29E-01	1.0470	1.58E+00	2.8549
GP_miner	0.5303681	0.3012	5.08E-01	0.9510	1.52E+00	2.7999
KN	0.5303681	0.5069	3.32E-01	0.5526	9.91E-01	1.3286
RBF_HM	0.5303681	0.4819	3.41E-01	0.5874	1.02E+00	1.3072
RBF_IM	0.5303681	0.4956	3.35E-01	0.5644	1.00E+00	1.3617
RBF_G	0.5303681	0.4976	3.34E-01	0.5626	9.99E-01	1.3391
RBF_DPS	0.5303681	0.4787	3.41E-01	0.5840	1.02E+00	1.3738
RBF_WCS	0.5303681	0.1437	4.51E-01	0.8867	1.35E+00	2.8033

TABLE 13.24

Error Metrics over In-bound and Relaxed Variable Bounds for Schittkowski Test Case 23

Case 23: x[0] * x[0] + x[1] * x[1] – (2.)

	R-Square Bound	R-Square Extended	RAAE Bound	RAAE Extended	RMAE Bound	RMAE Extended
AKR_G	1	0.9148	2.49E-15	0.2332	2.30E-14	0.9805
AKR_Exp	1	1.0000	7.01E-10	0.0001	1.82E-09	0.0004
KR_G	1	0.9993	7.18E-07	0.0185	4.09E-06	0.0739
KR_Exp	1	0.9042	9.35E-15	0.2588	3.03E-14	1.0779
GP_maxlik	1	0.7259	0.00E+00	0.6409	0.00E+00	2.1248
GP_miner	1	0.7524	1.34E-07	0.5391	4.68E-07	2.0492
KN	1	1.0000	6.86E-10	0.0001	1.78E-09	0.0004
RBF_HM	1	0.9951	3.42E-15	0.0464	1.90E-14	0.1951
RBF_IM	1	0.9992	1.29E-13	0.0165	3.77E-13	0.0864
RBF_G	1	0.9995	5.40E-14	0.0128	1.68E-13	0.0717
RBF_DPS	1	0.9947	2.19E-14	0.0413	6.52E-14	0.2279
RBF_WCS	1	0.6464	5.10E-15	0.5499	2.75E-14	2.4119

13.3.1 PERFORMANCE ANALYSIS FOR RESPONSE SURFACES SCHITTKOWSKI TEST CASES (2-D) 1–24

Error metrics of various response surfaces has been tabulated in Tables 13.3 to 13. 25. Response surfaces were evaluated on the basis of their prediction accuracy over data that are within the variable bounds, and data that are outside the variable bounds.

13.3.2 SUGGESTED SUPERVISED ML-BASED METHOD FOR VARIOUS SCHITTKOWSKI TEST CASES

From Tables 13.3 to 13.25, one can observe the performance of various supervised ML techniques with DEFAULT parameters. A similar analysis was performed for a few other test cases. Based on our expertise and experience, we have tabulated a list of suggested approaches that will be helpful in training models over a list of Schittkowski test cases. It must be noted that we have analyzed test cases where the number of variables varied between 2 and 6.

In real-life problems, the number of variables can vary. In this book, too, we have reported problems where we have analyzed data sets with 57 parameters. We will be presenting another case study with 100 variables where we developed models through the deep learning algorithm (Chapter 16). Thus, it is important for a data scientist to know the difference between various techniques as per the problem, especially while working with AI/ML-based predictive models. For a beginner, they

TABLE 13.25

Error Metrics over In-bound and Relaxed Variable Bounds for Schittkowski Test Case 24

Case 24: (pow(x[0] – 3., 2.) – 9.) * x[1] * x[1] * x[1] / (sqrt(3) * 27.) – (–1.)

	R-Square Bound	R-Square Extended	RAAE Bound	RAAE Extended	RMAE Bound	RMAE Extended
AKR_G	1	0.1175	1.34E-15	2.9797	4.92E-15	17.4021
AKR_Exp	0.9999991	0.9795	2.34E-04	0.0712	8.52E-04	0.4940
KR_G	1	0.9367	4.94E-07	0.1331	1.72E-06	1.1632
KR_Exp	1	0.4409	5.40E-15	0.9723	1.82E-14	5.9274
GP_maxlik	1	0.2976	5.78E-16	1.4858	2.79E-15	8.8441
GP_miner	1	0.3237	1.12E-09	1.3603	8.30E-09	8.5750
KN	1	0.9801	7.91E-13	0.0742	2.98E-12	0.5168
RBF_HM	1	0.7563	4.89E-16	0.4349	1.70E-15	2.4499
RBF_IM	1	0.9881	2.72E-12	0.0574	7.70E-12	0.3939
RBF_G	1	0.9655	1.17E-12	0.1045	3.01E-12	0.6876
RBF_DPS	1	0.9249	1.46E-13	0.1600	3.45E-13	1.0858
RBF_WCS	1	0.4712	4.42E-16	0.8447	1.84E-15	5.2982

can start with the one-dimensional and two-dimensional test cases in this chapter. Thus, working with different ML-based techniques is essential as real-world problems are different from Schittkowski test cases.

13.4   KNIME SOFTWARE: SCHITTKOWSKI TEST PROBLEM 47

KNIME software was used for developing models based on the random forest algorithm for Schittkowski test problem 47. The workflow developed in KNIME is shown in Figure 13.6, while model parameters have been tabulated in Table 13.27. In Figure 13.6, one can observe that several additional nodes exist on the workflow. This was done to demonstrate the software feature. At any instant, a user can perform a completely different analysis by just adding one node. A user does not need to write a single sine of computer program (code) if they do not want to code. Models developed in other platforms can be easily integrated with this workflow.

In Figure 13.6, one can observe that there are three nodes for Excel Reader. On one of them it is mentioned, training data, while on the other two it is mentioned in the testing set. The random forest model was trained on the data included in the node labeled "Training Data". "Testing set 1" contains data where the variable bounds for all the variables were relaxed by 5%. "Testing set 2" contains data where the variable bounds for all the variables were relaxed by 10%. Correlations were

TABLE 13.26

Suggested Supervised ML Algorithm for Various Schittkowski Test Cases

# Variables	Scitkowski Case	Best Performance
2	Case 1	AKR_G. KR_G
2	Case 2	AKR_G. KR_G
2	Case 3	AKR_G. KR_G
2	Case 4	AKR_G. KR_G
2	Case 5	RBF_HM. RBF_IM
2	Case 6	AKR_EXP. KN
2	Case 7	KR_G. KN. RBF_IM. RBF_DPS. RBF_WCS
2	Case 9	AKR_EXP. KN. RBF_IM
2	Case 10	AKR_EXP. KR_G. KN. RBF_HM. RBF_G
2	Case 11	AKR_EXP. KN
2	Case 12	AKR_EXP. KN
2	Case 13	AKR_EXP. KN
2	Case 14	AKR_EXP. KN
2	Case 15	AKR_EXP. KN
2	Case 16	AKR_EXP. KN
2	Case 17	AKR_EXP. KN
2	Case 18	AKR_EXP. KN
2	Case 19	AKR_EXP. KN
2	Case 20	AKR_EXP. KN
2	Case 21	AKR_EXP. KN
2	Case 23	AKR_EXP. KN
2	Case 24	KN. RBF_IM
3	Case 25	RBF_IM
3	Case 26	RBF_IM
3	Case 27	AKR_EXP. RBF_IM
3	Case 28	AKR_EXP
3	Case 29	AKR_G. AKR_EXP
3	Case 30	AKR_EXP

calculated between the actual data for Schittkowski test problem 47 and the random forest model prediction over the two testing sets. Correlations values have been tabulated in Table 13.28.

From Table 13.28, one can observe that the correlation values decrease with an increase in relaxation of variable bounds. This behavior is expected. The comparison between Schittkowski test problem 47 data and random forest model prediction has been plotted in Figures 13.7 (in bound), 13.8 (5%), and 13.9 (10%).

We collaborated with KNIME official Mr. Scott Fincher. He provided us with vital information regarding workflow development in KNIME software. The workflow developed in collaboration with KNIME is shown in Figure 13.10. This

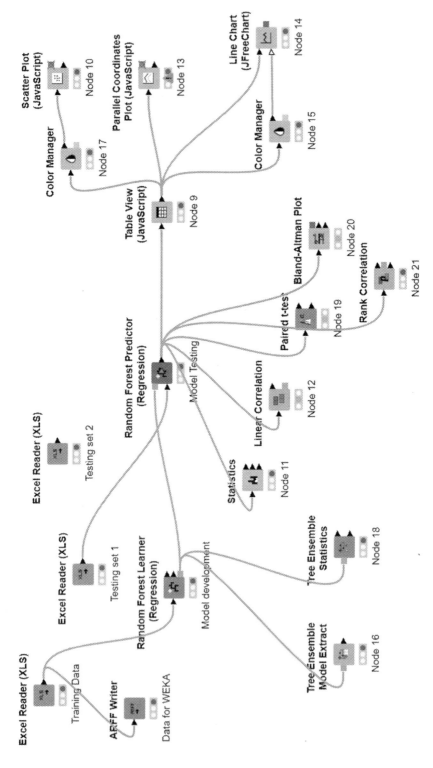

FIGURE 13.6 Workflow for Schittkowski test problem (case) 47.

TABLE 13.27

KNIME Software: Model Parameters (Random Forest) for Schittkowski Test Problem (Case) 47

Model Parameters (Random Forest)	Values
Number of models	100
Minimal depth	10
Maximal depth	17
Average depth	12.27
Minimal number of nodes	105
Maximal number of nodes	131
Average number of nodes	116.36

TABLE 13.28

KNIME Schittkowski Test Problem 47: Correlation between Actual Data and Model Predictions over the Training Set and the Two Testing Sets

	Within Bound	5%	10%
Correlation	0.965	0.792	0.768

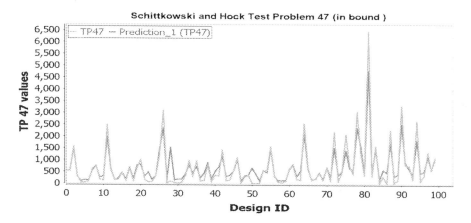

FIGURE 13.7 KNIME Schittkowski test problem 47: Correlation between actual data and model predictions over data which is within variable bounds.

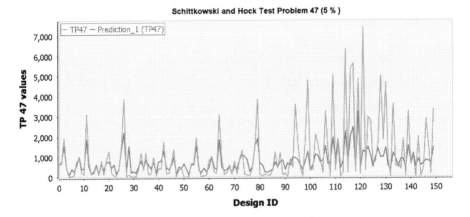

FIGURE 13.8 KNIME Schittkowski test problem 47: Correlation between actual data and model predictions over the testing set 1(5%).

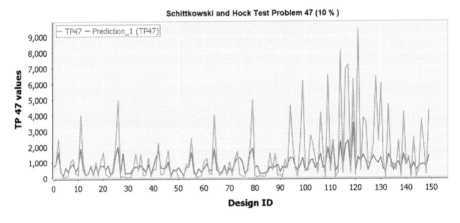

FIGURE 13.9 KNIME Schittkowski test problem 47: Correlation between actual data and model predictions over the testing set 2 (10%).

workflow is cleaner when compared with Figure 13.6, as there are only those nodes that are needed for this work. In this workflow (Figure 13.10), the training and testing set were divided by a partitioning node from the same Excel Reader node.

13.5 CONCLUSIONS

The purpose of this test case is to demonstrate the importance of parameter tuning. All of these models' performance can be significantly improved. The accuracy of a model prediction depends on several factors like data preprocessing, removing outliers, scaling, and tuning of parameters. There are platforms where a user may upload the data and it generates an accurate model, or the platform provides a model with significant error or uncertainty. A user must not make conclusions about an algorithm or a computational platform or software on the basis of just a few cases.

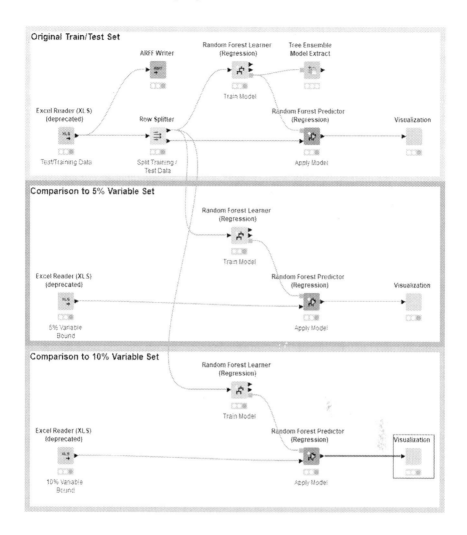

FIGURE 13.10 KNIME: Workflow developed in collaboration with KNIME official (Mr. Scott Fincher).

For 2-D cases, we have listed a set of equations that performed better outside the variable bounds.

In KNIME software, we have used just one algorithm. We used the default parameter to demonstrate that even with the default parameters, one can get nice fit.

For predicting outside the variable bounds, one must work on parametric tuning prior to drawing conclusions or making any general statement about any computational platform, software or an algorithm.

14 Case Study #14: Effect of Operating Parameters on Roll Force and Torque in an Industrial Rolling Mill: Supervised and Unsupervised Machine Learning Approach

14.1 INTRODUCTION: ROLLING DATA

Rolling data used in this case study was acquired from an industrial rolling mill. The data sheet was analyzed and relevant information was tabulated, as seen in Table 14.1. Among the listed operational parameters in Table 14.1, accurate estimation of roll force and roll torque is critical for an operator. Fluctuation in roll force and torque directly affect the quality of the product. In an industrial environment, there exist several parameters that can affect the roll force and torque.

In Table 14.1, we have tabulated a list of parameters that were acquired from an operational rolling mill. Regarding parameters, it totally depends on the operator of the rolling mill on what parameters they want to work with and their final objectives. In a previous chapter on blast furnaces, we explained this in detail. In that chapter, operational data from three blast furnaces from an integrated steel plant was studied. All three furnaces produce hot metal (molten iron) and slag as their final products, but the operating parameters listed for the three furnaces were different and the number of operating parameters varied. Additionally, desired objectives varied for these three furnaces.

A rolling mill is an integral part of an integrated steel plant. In this book, we have presented case studies for three blast furnaces from three different integrated steel plants and also included a case study on a BOF/LD furnace from an integrated steel plant. Thus, the rolling parameter an operator chooses for a particular study or the objective can be different for different rolling mills from different integrated steel plants. From application point of view, bulk properties of importance include yield strength (YS), ultimate tensile strength (UTS) and elongation percent. In this chapter, these properties have been optimized in the latter part with respect to operational parameters in a rolling and coiling mill.

DOI: 10.1201/9781003167372-14

TABLE 14.1
Rolling Parameters

	Rolling Parameters	Notations in modeFRONTIER	Unit	Minimum	Maximum
1	Reheating time	Reheat_T	minutes	135	1188
2	Roll diameter	Roll_Dia	mm	1042.705	1120.265
3	Roll crown	Roll_Cr	μm	0.3	216.65
4	Entry thickness	Ent_Thk		15.2	250
5	Width	Width	mm	1613.43	4181.9
6	Length	Length		1794	38139
7	Temperature	Temp	°C	889	1233
8	Speed	Speed	m/s	1.5	5.6
9	Wait time	Wait_T	seconds	6	225
10	Reduction	Reduction	–	0.02	0.38
11	Strain	Strain	–	0.021491	0.554531
12	Strain rate	St_Rt	s^{-1}	0.573413	16.24724
13	Flow stress	Fl_Stress	Mpa	22.5184	163.1113
14	Roll force	Force	KN	7899	59433
15	Roll torque	Torque	KN.m	683	5968

In this case, operators worked on estimation of roll force and roll torque. From Table 14.1, operational parameters with serial numbers between 1 and 13 are expected to affect the roll force and torque for this present case. An operator has control over parameter with serial number between 1 to 9 only. The correlation matrix for the operational parameters and targeted objectives have been shown in Figure 14.1.

In Figure 14.1, one can observe that several parameters and objectives are strongly correlated. But, for the majority of the table, the correlation coefficient is extremely low. Operational parameters like roll crown, roll diameter, and reheat time are not correlated with any parameters or objectives. In an industrial setup, such weak correlation is expected due to the complexity of the manufacturing process.

In an industrial environment, there exist several parameters that can affect the final objective. Additionally, operational parameters are interrelated; that is, one parameter can affect another parameter. Since all the parameters are expected to affect the final objective, it is important to study correlations between the operating parameters in addition to studying the effect of operating parameters on the final objective.

Thus, we processed this data for analyzing patterns in the data set using an unsupervised machine learning approach: self-organizing maps (SOMs).

14.2 UNSUPERVISED MACHINE LEARNING: SELF-ORGANIZING MAPS (SOMS)

In SOM analysis, X-Dimension = 12, Y-Dimension = 13; thus, total map units are equal to 156. The neighbor's function is Gaussian. Logistic scaling was performed for minimizing error. The quantization error is equal to 0.083, while the topological error is equal to 0.084. The SOM for rolling data is shown in Figure 14.2.

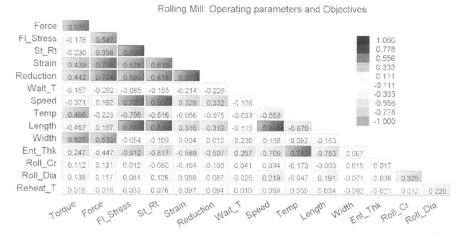

FIGURE 14.1 Correlation matrix: Rolling parameters and objectives.

In Figure 14.2, one can observe that several operational parameters and objectives are grouped together on the component plot. In order to distinguish different units on the SOM plot, we have added serial numbers corresponding to each parameter from Table 14.1. Three operational parameters are isolated: roll diameter, roll crown, and entry thickness. It can be understood that entry thickness values are fixed for most of the slabs. Roll diameter and roll crown values are fixed for several cases of rolling. Thus, there is less variation in the values for entry thickness, roll crown, and roll diameter, so it may not be correlated with other operational parameters and objectives. These values can be used as classes for developing predictive models for roll force and torque, but then the amount of data will significantly decrease.

In Figure 14.2, reduction and strain are grouped in between force and torque. Figure 14.3 shows the comparison between these units for better understanding.

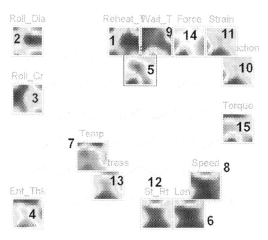

FIGURE 14.2 SOM component plots for various parameters and objectives identified in Table 14.1.

In Figure 14.3, one can observe that the distribution of strain and reduction is almost identical in the SOM space. That means the higher the strain, the higher will be the reduction, which an operator knows, and the SOM plots also show that. Thus, SOM plots can pick correlations that are well understood and can be explained to the operator. Hexagonal unit cell for maximum force, reduction, and strain are identical. Torque values are not the highest for this particular unit. Thus, the highest force combined with a high value of torque will result in high strain values and the highest reduction. This correlation can again be understood from the operator's point of view. Additionally, the region of lowest force, torque, reduction, and strain are in the same neighborhood. This means that the lowest values of force and torque will result in lower straining and thus the lowest reduction. Thus, SOMs can be helpful in determining correlations from a complex data set obtained from the rolling mills.

Figure 14.4 shows the distribution of a few other operational parameters along with the reduction unit, which was already analyzed in Figure 14.3. Reduction unit was

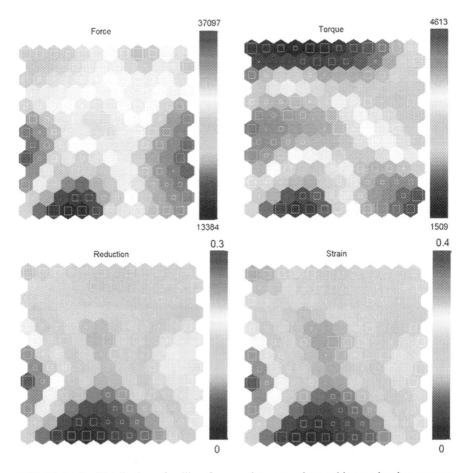

FIGURE 14.3 Distribution of rolling force and torque, along with correlated parameter reduction and strain.

included in this analysis because there are some patterns shown in the reduction unit that match the patterns shown in the following units: speed, strain rate, flow stress, and length. The only difference is that the highest value for the reduction is in a different region. The patterns shown on speed, strain rate, flow stress, and length are almost identical, while the pattern shown for temperature is almost opposite to speed, strain rate, flow stress, and length. For the highest values of speed, strain rate, flow stress, and length, reductions are average, because the temperature is lowest for these regions on the SOM plot. Temperature is highest for the lowest values of speed, strain rate, flow stress, and length, and for these few units, reduction is the lowest. These correlations are understandable as reduction is affected by strain rate, flow stress, and velocity or speed.

The SOM units for roll diameter, roll crown, reheat time, and force have been analyzed in Figure 14.5. Roll diameter and reheat time correlate to some extent. The highest and lowest value regions for both reheat time and roll diameter is in the same region on the SOM plots. The unit containing the highest value for roll diameter, roll crown, and reheat time are identical (top-right corner). Roll force seems to have some weak correlation with other parameters on the figure, but it is in the extrapolated region; that is, the unit does not include actual data from the rolling mill.

All the recommendations, suggestions, or correlations reported in this chapter for SOM analysis are for the units that contain actual data from the rolling process. Extrapolated region can be trusted as SOM analysis is known for preserving the topology of the data. Again, it is on the operator to decide. Through SOM analysis, we can provide suggestions/recommendations, but the operator is responsible for maintaining the desired quality of the product.

SOM units for entry thickness, width, wait time, force, torque, and strain rate have been analyzed in Figure 14.6.

In Figure 14.6, strain rate is opposite to entry thickness. We have already discussed the correlation of strain rate with other operating parameters. Thus, all those relations will hold. Units in the region of highest entry thickness have the lowest strain rate, lowest width, lowest roll force, and lowest torque. The lowest entry thickness combined with average width corresponds to the lowest torque and average force.

In SOM analysis, one can notice the presence of a square inscribed in the hexagonal units. The SOM algorithm has been discussed in detail in Chapter 2 and in other chapters through case studies. On a SOM, candidates (rows in a data set) are positioned over the vertices of the hexagonal units. The more numbers of candidates associated with a hexagonal unit cell, the larger will be the square. Through SOM analysis, an operator or a data scientist can identify the hexagonal units which provide them with optimum roll force and torque. Then they can click on that unit and identify the candidate (rows ID) that are positioned on the edges of that unit. The most important objective of an operator is to find a balanced set of operating parameters that will provide them a final product with the desired quality. In an industrial environment, lots of operational parameters affect the quality of the final product, some of which an operator has no control over. An operator has to deal with any adverse challenge that may arise due to any irregularities in the process parameter or chemistry of the product. Thus, it is important for an operator to know the complex correlations between various operating parameters and the desired final

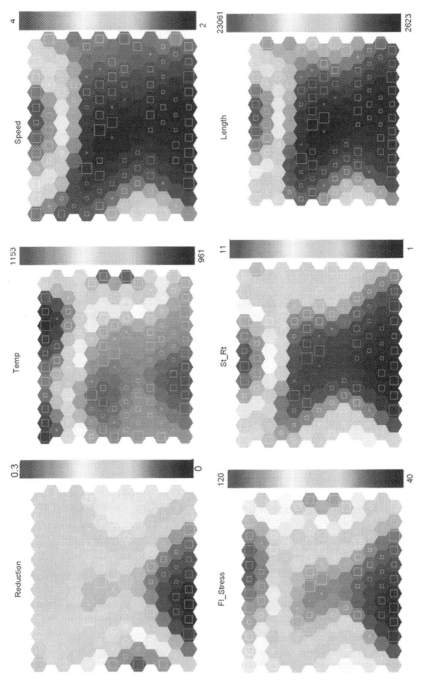

FIGURE 14.4 SOM analysis of operating parameters for a rolling mill.

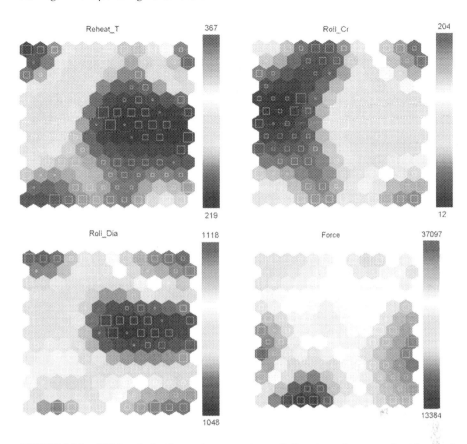

FIGURE 14.5 SOM analysis of operation parameters: roll diameter, roll crown, reheat time, and force.

objective, that is the quality of the product. SOM analysis can be helpful as even a small set of data can be efficiently analyzed through SOM analysis.

SOM analysis was important for this type of data set that we obtained from the rolling mill. The SOM algorithm is an unsupervised machine learning algorithm. It uses all the data and various patterns are observed. Most of these correlations can be established through literature. In this rolling data, there are lots of parameters that are dependent on each other. For example, an operator can choose reduction as an objective and also as a parameter; that is, an operator can define reduction as an objective, and make an attempt to find the best set of parameters that will help them achieve reduction of that order. Through SOM analysis, a user has the liberty of treating all the components of the SOM component plot the way they want. They can treat it as an operational parameter or an objective. This will be helpful for them in determining which parameters can be chosen as variables when they proceed towards developing predictive models through supervised machine learning.

FIGURE 14.6 SOM analysis of operation parameters: entry thickness, width, wait time, force, torque, and strain rate.

14.3 SUPERVISED MACHINE LEARNING: K-NEAREST NEIGHBOR ALGORITHM IN modeFRONTIER

14.3.1 Variable Screening

Roll force and roll torque were identified as objectives. The rest of the parameters in Table 14.1 were identified as variables that affect roll force and roll torque. Variable screening was performed through the smoothening ANOVA algorithm in mode-FRONTIER. Figure 14.7 shows the plot for roll force and torque. In Figure 14.7, one can observe that strain has the highest contribution index for both roll force and torque. In other chapters, we have provided a table for variable screening. In this chapter, we have provided the plots.

14.3.2 Predictive Models Developed Through Supervised Machine Learning

In modeFRONTIER, the K-NN algorithm was used for developing the regression models for roll force and roll torque as a function of operating parameters

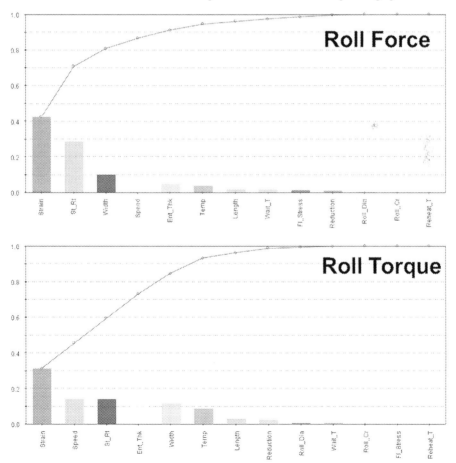

FIGURE 14.7 Effect of operational parameters on roll force and torque.

mentioned in Table 14.1. Figure 14.8 shows the comparison between industrial data and the K-NN model prediction for roll force. Figure 14.9 shows the comparison between industrial data and the K-NN model prediction for roll torque.

Model predictions are acceptable for both roll force and torque. We are dealing with data from the industry. Industrial data are affected by several factors, and as per our experience with industrial data, one must avoid overfitting. If the pattern seems correct, models can be trusted. An operator takes into account predictions from automation models in the industry, but they always have the liberty of making the final decision when it comes to applying the model predictions as their job is to ensure that the quality of the product is acceptable.

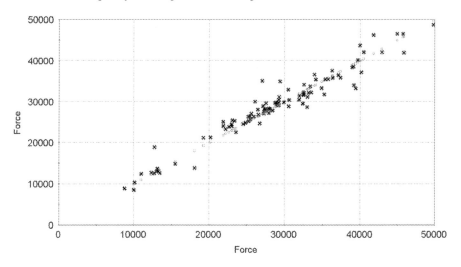

FIGURE 14.8   Roll force: Industrial data vs. K-NN model prediction.

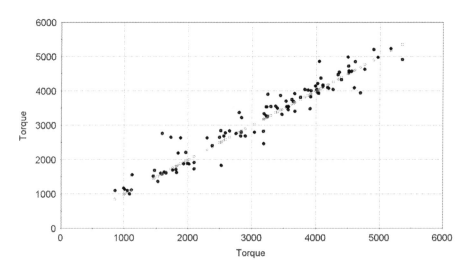

FIGURE 14.9   Roll torque: Industrial data vs. K-NN model prediction.

In this book, we used the K-NN algorithm a lot in modeFRONTIER. The reason is that modeFRONTIER is a proprietary software, but the K-NN algorithm is quite simple and available in lots of computational platforms. If a reader wants to perform similar work, they can use other computational platforms and use this algorithm for free.

In the next section, we have used an open-source computational platform, KNIME. In KNIME, a user can develop complex models and perform complex statistical calculations without writing a single line of code. A user can also use their own computer code, or models developed from different computational platforms in KNIME.

14.4 KNIME: RANDOM FOREST ALGORITHM

In KNIME software, we developed predictive models for roll force and torque as a function of parameters mentioned in Table 14.1. The workflow developed in KNIME software for this case study is shown in Figure 14.10. For developing prediction models, we chose the random forest algorithm for developing models. The K-NN algorithm is also available in KNIME software.

The prediction accuracy of a model depends on tuning the parameters. Comparing the same algorithm in two different platforms does not fulfill any purpose. Thus, we chose the random forest algorithm in KNIME.

In Figure 14.10, the workflow is quite simple. In the beginning, a user needs to arrange data in an Excel sheet. In the KNIME platform, they can choose the "Excel Reader (XLS)" node and they have to configure this node in order to format the table as per requirements of the KNIME software. Thereafter, they can choose the nodes shown in Figure 14.10, configure the nodes, and execute the nodes. They can execute one node at a time to work in steps or execute the entire workflow at the same time. The entire process is quite fast, and the entire work reported in this section was performed in a few minutes.

In Figure 14.10, it can be seen that there is a "Partitioning" node. This node was used for dividing data into training (89%) and testing (11%) sets. The node marked as "Random Forest Learner (Regression)" uses the training set for developing the models for roll force and torque, while another node, "Random Forest Predictor (Regression)", uses the testing set data for testing the model predictions. Correlation between the actual plant data and random forest model predictions were calculated through the node labeled "Linear Correlation". The node marked "Line Chart (JFree Chart)" is used for plotting.

Figure 14.11 shows the comparison between the actual production data from the rolling mill for roll force and the random forest model predictions for rolling force. One can observe that random forest models were able to mimic the trends shown in the data set for rolling force. The correlation coefficient was calculated between actual plant data and the model predictions for roll force; the correlation value is 0.987. Thus, a random forest model developed in KNIME software can be used as a predictive tool for estimation of rolling force.

Figure 14.12 shows the comparison between the actual production data from the rolling mill for roll torque and the random forest model predictions for rolling torque. One can observe that random forest models were able to mimic the trends shown in the data set for rolling torque. The correlation coefficient was calculated between actual plant data and the model predictions for roll torque; the correlation

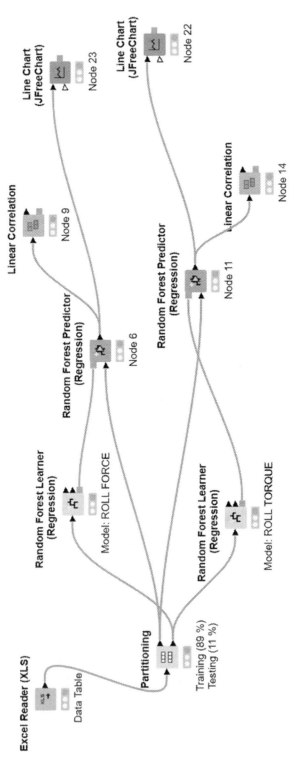

FIGURE 14.10 Workflow developed in KNIME software for this case study.

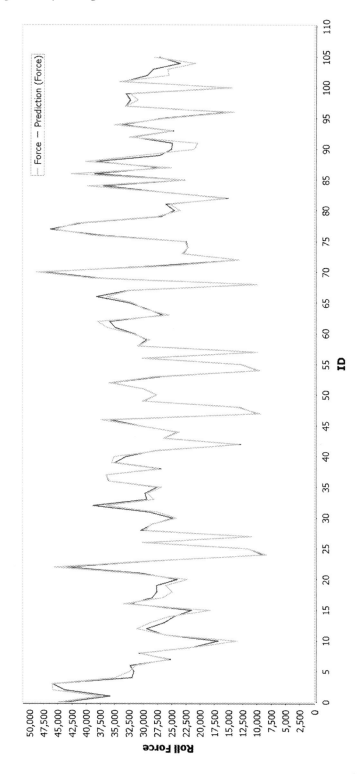

FIGURE 14.11 Roll force: Comparison between actual plant data and random forest model predictions in KNIME.

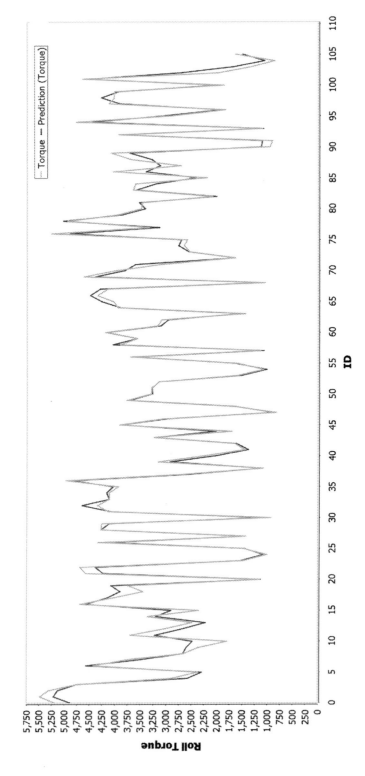

FIGURE 14.12 Roll torque: Comparison between actual plant data and random forest model predictions in KNIME.

value is 0.984. Thus, a random forest model developed in KNIME software can be used as a predictive tool for estimation of rolling torque.

It always depends on the data scientist on what approach they want to consider for a particular analysis. Even for model parameters, a data scientist can decide whether they want to tune the parameters, or whether they want to go with the "Default" settings. For most of the computational platforms that we have presented in this book, to start with, one can even work with the default settings or parameters. Then they can work on hyper-tuning model parameters.

14.5 HIERARCHICAL CLUSTERING ANALYSIS (HCA)

HCA analysis was performed to study correlations within the data set. The data set was divided into eight clusters. HCA parameters are tabulated in Table 14.2. The HCA algorithm has been discussed in detail in Chapter 3. All of these parameters have been explained in Chapter 3. In this chapter, we have discussed a few important correlations observed in the data set.

Figure 14.13 shows the clustering plot for roll force versus torque. Apart from cluster 7, all the other clusters are almost aligned in the same direction. This indicates that for this case of rolling mill, force and toque are directly related to each other. An increase in roll force will lead to an increase in torque. From Figure 14.13, an operator can check the data included in cluster 7 for irregularities, as it follows a different pattern when compared with the rest of the clusters.

Figure 14.14 shows the clustering plot for roll force versus reduction. One can observe that for six of the clusters, an increase in roll force will lead to an increase in reduction. In Figure 14.13, cluster 7 was oriented in a different manner. But in Figure 14.14, it can be observed that cluster 7 is aligned with other clusters. Two of the clusters are almost flat and their ellipse is larger in size. This means that for the candidates in these clusters, there is no correlation between roll force and reduction.

Figure 14.15 shows the clustering plot for roll torque versus reduction. For four of the clusters, an increase in torque will lead to an increase in reduction. Three of

TABLE 14.2
Parameters for HCA Analysis

Cluster	Size	ISim	ESim
0	141	1.4	0.8
1	48	1.4	0.7
2	100	1.3	0.7
3	109	1.5	0.7
4	231	1.6	0.9
5	185	1.9	0.9
6	91	1.4	0.7
7	65	1.4	0.8

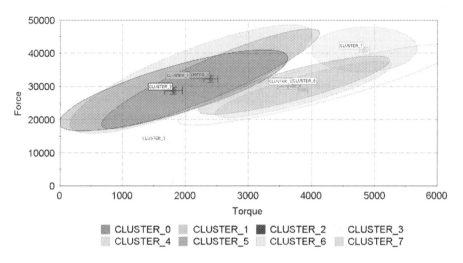

FIGURE 14.13 HCA analysis: Roll force vs. torque.

the clusters are flat, so it is a mixed correlation or no correlation. For cluster 7, an increase in torque will lead to a slight decrease in reduction. Thus, cluster 7 can be analyzed by an operator as it is affecting the torque. Parameters of the candidates included in this cluster can be analyzed for irregularities.

Figure 14.16 shows the clustering analysis for reduction and strain rate. Two of the clusters are completely flat and one is almost flat, meaning no to mixed affect. The rest of the clusters are oriented in a way that an increase in strain rate will lead to an increase in reduction.

In Figure 14.17, for clusters 0, 1, 2, 3, and 7, an increase in flow stress will lead to an increase in force. For three other clusters, we cannot clearly state anything as the clusters are parallel to the flow stress axis.

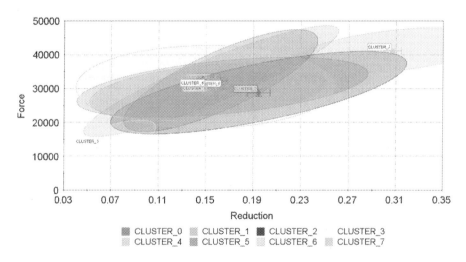

FIGURE 14.14 HCA analysis: Roll force vs. reduction.

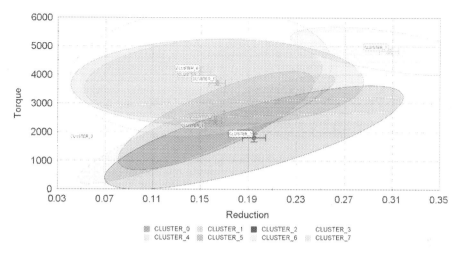

FIGURE 14.15 HCA analysis: Roll torque vs. reduction.

In Figure 14.18, one can observe that for two values of roll diameter, there are small clusters where force suddenly increases with roll diameter. For the rest of the clusters, there is no such correlation.

In Figure 14.19, for cluster 6, roll diameter increases with roll crown, while for cluster 0, roll diameter decreases with an increase in roll crown.

As we have mentioned in SOM analysis, one can treat both roll diameter and roll crown as classes if they have a large database. In this work, supervised models have acceptable accuracy even when we included the roll diameter and roll crown as variables. Therefore, it depends on the quantity of data we have.

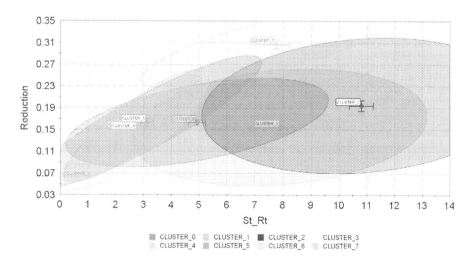

FIGURE 14.16 HCA analysis: Reduction vs. strain rate.

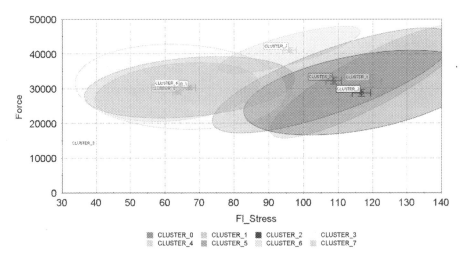

FIGURE 14.17   HCA analysis: Roll force vs. flow stress.

The industrial rolling mill is a major installation that depends on many factors. Through HCA analysis, data over a period can be divided and analyzed efficiently. An industrial process is prone to breakdown, which can be caused by a certain parameter. Through HCA analysis, the effect of various parameters can be analyzed. Then an operator can easily determine if there is any irregularity due to a specific parameter or if a certain parameter is helping them in increasing efficiency of the entire rolling process. Through clustering analysis, an operator can monitor the entire rolling process by analyzing a few clustering plots as they already know the entire setup well. All they need from time to time is to monitor the process for a smooth process and avoid breakdowns. Operators use automation models for

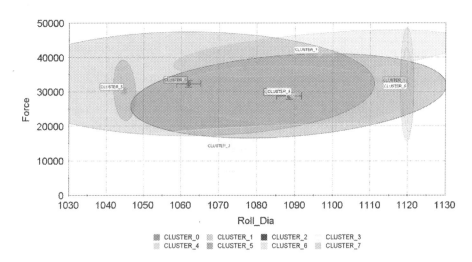

FIGURE 14.18   HCA analysis: Force vs. roll diameter.

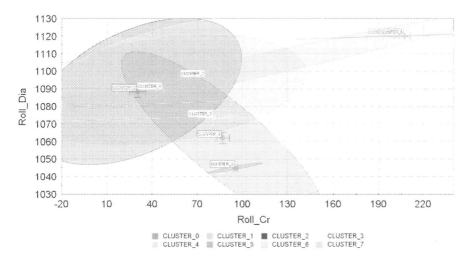

FIGURE 14.19   HCA analysis: Roll diameter vs. roll crown.

reference from time to time. Thus, this clustering analysis will help them in even better understanding the entire process.

In the next section, we have performed a principal component analysis (PCA) on the clustering data.

14.6   PRINCIPAL COMPONENT ANALYSIS

Figure 14.20 shows the distribution of data obtained from the rolling mill over the PC1 and PC2 spaces. The color shown in the figure is the same color associated

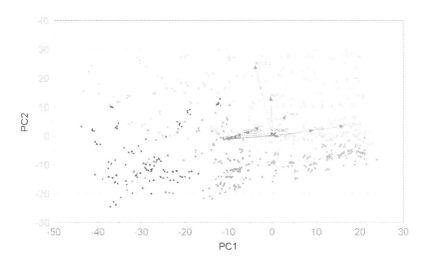

FIGURE 14.20   PCA analysis: PC1 and PC2.

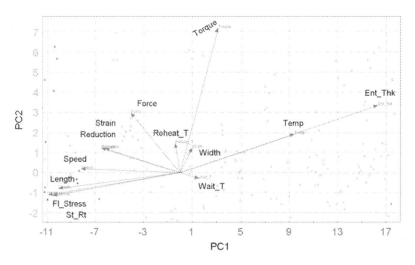

FIGURE 14.21   Figure 14.20 Enlarged view for PCA analysis: PC1 and PC2.

with the clusters in the HCA analysis. Arrows are difficult to visualize in Figure 14.20; thus, an enlarged view is presented in Figure 14.21.

In Figure 14.21, one can observe a few correlations that match the SOM plots. On the SOM plots, units for strain rate, length, speed and flow stress were close to each other. In Figure 14.21, the arrows for these units are together and about the same length. Strain and reduction are clustered together on the SOM plot (Figure 14.2) and in PCA plot one can observe the same correlation.

In the SOM component plot in Figure 14.2, on one side of the force, one can observe strain and reduction, while on the other side of force one can observe wait time, reheat time, and width. In the PCA plot, one can observe that on one side of the arrow for force, there are strain and reduction, while on the other side of force, there is wait time, reheat time, and width.

It is extremely important that some of the correlations match even if the algorithm is different. New patterns can be observed and must be investigated. But, the process is same, so whatever the algorithm, process parameters must affect the desired objective in somewhat the same manner. Quantitatively, one must not expect the same level of accuracy in two different algorithms, but some patterns must match in all the data analysis.

14.7   PLATE MILL: MULTI-OBJECTIVE OPTIMIZATION OF YS, UTS AND ELONGATION PERCENT OF MICRO-ALLOYED STEEL PLATES

From application point of view, three properties were identified: Yield strength (YS), ultimate tensile strength (UTS) and elongation percent. In this section, we utilized operational data from the same plate mill (section 14.1 to 14.6). But, we used new operational parameters and included the composition of the plates as a parameter that affects the targeted properties. Operational parameters used for this case study are listed in

Table 14.3 In the log sheet obtained from industry, there were several observations corresponding to Charpy test, both at the room temperature and at sub zero temperature (−20°C). In this section, we did not include the readings from the Charpy test as our objective was to maximize the three desired properties: YS, UTS and elongation percent. In Table 14.3, carbon equivalent (CE) is also included. Carbon equivalent is an indicator of susceptibility to cracking and is calculated on the basis of composition of steel. Microalloyed steel containing elements like Nb, V, B, are prone to cracking in the temperature range of 700 to 900°C, thus CE values were included in this table. In this section, we have not utilized CE in data analysis. In Chapter 12, we have discussed crack

TABLE 14.3

Plate Mill: Operational Parameters along with Their Notation, Units and Variable Bounds

	Parameter/objective	Notation	Units	Min.	Max.
1	Thickness	Thk	mm	12	63
2	Weight	Wt	tonne	1.731	14.837
3	YS	YS	Mpa	353	493
4	UTS	UTS	Mpa	496	632
5	Elongation (EL5.6)	El_per	%	22	30
6	Discharge temperature	Dis_T	°C	1163	1262
7	Heating time	H_time	Hours	3.57	18.36
8	P1 temperature	P1_T	°C	1011.14	1170.9
9	Last pass temperature	LP_T	°C	782.01	1094.2
10	Final temperature	F_T	°C	720.74	974.5
11	Last rough reduction	L_R_Red_per	%	12.84	42.7
12	Number of passes	N_Pass		11	17
13	First fine reduction	F_F_Red_per	%	15.17	32.07
14	Final reduction	F_Red_per	%	6.03	25.73
15	Waiting thick	W_thk	mm	19.73	99.28
16	Restart temperature	Rstrt_T	°C	914.01	1113.37
17	Start temperature	Start_T	°C	1060.08	1178.13
18	Wait time	W_time	Seconds	0	432.5
19	C	C	weight %	0.15	0.2
20	Mn	Mn	weight %	1.02	1.42
21	P	P	weight %	0.013	0.027
22	S	S	weight %	0.002	0.026
23	Si	Si	weight %	0.167	0.35
24	Al	Al	weight %	0.011	0.049
25	Nb	Nb	weight %	0.002	0.055
26	V	V	weight %	0	0.05
27	Cu	Cu	weight %	0.001	0.005
28	B	B	weight %	0.001	5
29	Carbon equivalent (CE)	CE	%	0.348	0.444

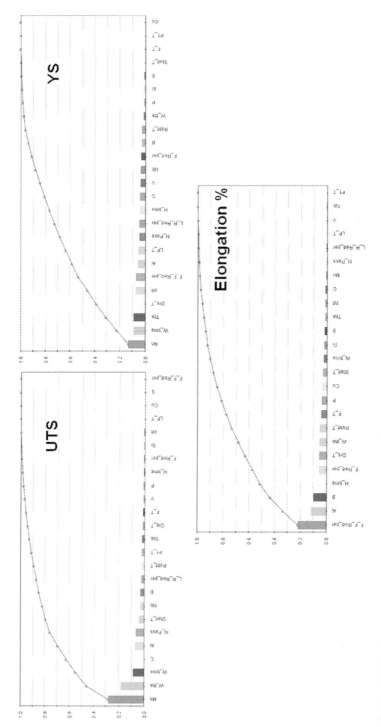

FIGURE 14.22 Statistical analysis on plate mill data: Effect of operating parameters and composition on the desired properties: YS, UTS and elongation percent.

susceptibility coefficient (CSC) for microalloyed steel within the framework of CALPHAD approach. Thus, a discussion on role of CE has not been included in this chapter.

In this section, we have focussed on the statistical analysis and multi-objective optimization. A total of 440 plates was included in the log sheet. Upon data-preprocessing, we ended up with about 400 plates. There were several missing points in the datasheet that needs to be avoided. Variable screening was performed through the smoothening ANOVA algorithm in modeFRONTIER. Figure 14.22 shows the effect of various operational parameters including composition on the three targeted properties. In Figure 14.22, with respect to the contribution index, manganese is the highest contributor among all variables for both YS and UTS. From the metallurgical point of view, manganese is a major alloying element for various grades of high strength steel, like high-strength low alloy (HSLA), advanced high-strength steel (AHSS) grades etc. Meanwhile, elongation percent is affected by a reduction in the final stage.

Prediction models for the targeted properties were developed through the K-NN algorithm. Models were developed in modeFRONTIER software. It was followed by multi-objective optimization of targeted properties. Evolutionary or genetic algorithm (EA/GA) chosen for this case were: multi-objective particle swarm optimization (MOPSO), and non-dominating sorting genetic algorithm II (NSGA2). Optimum results were obtained through NSGA2 algorithm. Figure 14.23 shows the Pareto-front obtained through the NSGA2 algorithm for the plate mill.

In Figure 14.23, each point on the figure corresponds to a new plate. Properties like YS, UTS and elongation percent, can be ascertained from the corresponding axis. These are the optimum values, and for each plate there is a list of operating parameter along with the composition which will be helpful in achieving these optimum values.

We analyzed data from another coiling mill and the optimization results are shown in the next section.

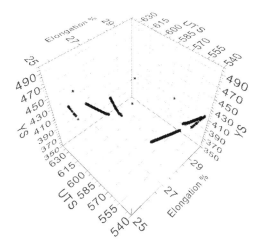

FIGURE 14.23   Plate mill: Pareto-front obtained after multi-objective optimization of targeted properties: YS, UTS and elongation percent.

14.8   COILING: MULTI-OBJECTIVE OPTIMIZATION OF YS, UTS AND ELONGATION PERCENT OF AHSS COILS

Operational data from a coiling mill was analyzed through statistical tools. Thereafter, multi-objective optimization of desired properties were performed to determine operational parameters and composition that will be helpful in achieving these optimized properties. From application point of view, three properties were identified: Yield strength (YS), ultimate tensile strength (UTS) and elongation percent. Hardness has been included in the statistical analysis, but was not included in multi-objective optimization. Table 14.4 shows the list of operational parameters along with variable bounds and notations for the coiling mill. There were few values of carbon equivalents, but it has not been included in this analysis.

TABLE 14.4

Coiling Mill: Operational Parameters along with Units, Notations, and Variable Bounds

Sl.No.	Parameter/objective	Notation	Units	Minimum	Maximum
1	Width	Width	mm	1025	1500
2	Thickness	Thkns	mm	7	10
3	Grain size of ferrite	GS	μm (micron)	9.7	11.6
4	C	C_p	Weight %	0.045	0.073
5	Mn	Mn_p	Weight %	0.132	1.54
6	Si	Si_p	Weight %	0.003	0.371
7	P	P_p	Weight %	0.006	0.14
8	S	S_p	Weight %	0.002	0.008
9	Al	Al_p	Weight %	0.014	0.057
10	N	N_p	Weight %	0.0044	0.012
11	Cr	Cr_p	Weight %	0.005	0.254
12	Ni	Ni_p	Weight %	0.004	0.06
13	Cu	Cu_p	Weight %	0.002	0.033
14	Nb	Nb_p	Weight %	0	0.05
15	V	V_p	Weight %	0	0.42
16	Ti	Ti_p	Weight %	0	0.047
17	Mo	Mo_p	Weight %	0	0.083
18	YS	YS	MPa	450	536
19	UTS	UTS	MPa	508	616
20	Hardness	Hardness		80	89
21	Elongation percent	El_Prct	%	23	40
22	Average finishing temperature	FT_Avg	°C	830.3	905.1
23	Average coiling temperature	CT_Avg	°C	451.6	614.1
24	Shear temperature	Shear_T	°C	644	1115
25	Casting speed	Cast_Speed	m/s	3.98	6.69

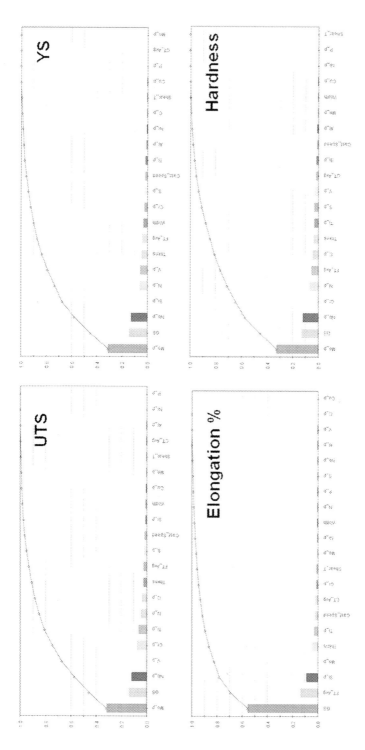

FIGURE 14.24 Statistical analysis on coil mill data: Effect of various operational parameters on the desired properties: yield strength (YS), ultimate tensile strength (UTS), elongation percent and hardness.

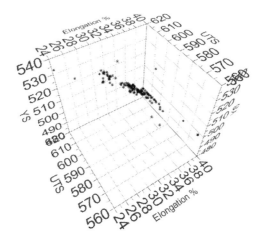

FIGURE 14.25   Coil mill: Pareto-front obtained after multi-objective optimization of targeted properties: YS, UTS and elongation percent.

A total of 250 coils was included in the log sheet. Upon data-preprocessing, we ended up with about 220 coils. There were several missing points in the datasheet that needs to be avoided. Variable screening was performed through the smoothening ANOVA algorithm in modeFRONTIER. Figure 14.24 shows the effect of various operational parameters including composition on the three targeted properties.

Prediction models for the targeted properties were developed through the K-NN algorithm. Models were developed in mode-FRONTIER software. It was followed by multi-objective optimization. Evolutionary or genetic algorithm (EA/GA) chosen for this case was multi-objective particle swarm optimization (MOPSO). Optimum results were obtained through MOPSO algorithm. Figure 14.25 shows the Pareto-front obtained through the MOPSO algorithm for the coils.

In Figure 14.25, each point corresponds to a separate coil with properties that can be ascertained from the corresponding axis of yield strength (YS), ultimate tensile strength (UTS), and elongation percent. For each point, there is a list of operational parameters that will be helpful in achieving those properties.

14.9   CONCLUSIONS

In this case study, we analyzed the data from an industrial rolling mill through a set of supervised and unsupervised machine learning algorithms using different software packages. Unsupervised machine learning algorithms were able to determine patterns within the data set that can be verified from literature and is understandable. Correlations observed through different approaches are similar in nature.

Supervised machine learning: Model accuracy is acceptable for the models developed using K-NN in modeFRONTIER and the random forest algorithm in KNIME. These models can be used as a predictive tool for the future.

Multi-objective optimization: Evolutionary or genetic algorithms (EA/GA) can be efficiently used for determining operational parameters, compositions, etc that will be helpful in manufacturing plate and coils with optimum properties. From application point of view, it is important to work on improving multiple properties of a given steel. Through genetic algorithms, a user can work on optimizing multiple properties of a given alloy and determine operational parameters that will be helpful in achieving those properties.

15 Case Study #15: Developing Predictive Models for Flow Stress by Utilizing Experimental Data Generated from Gleeble Testing Machine: Combined Experimental-Supervised Machine Learning Approach

15.1 INTRODUCTION: THERMOMECHANICAL DATA FROM GLEEBLE TESTING MACHINE

Alloying elements of a grade of steel include C, Si, Mn, P, S, Al, Nb, and Ti. This sample was tested in a Gleeble-3500 testing machine. Temperature and strain rate were varied for each test. Temperature values were fixed for each of the tests at 800, 900, 1000, and 1100°C, while the strain rate varied from 1 (/s), 10, 50, and 100/s. Machine output was in the form of data for a stress-strain curve. From this analysis, a significant amount of data was collected. Figure 15.1 shows the plots for the stress-strain curve (experiments) for various strain rates and temperatures.

15.2 SUPERVISED MACHINE LEARNING: DEVELOPMENT OF PREDICTIVE MODEL FOR STRESS

15.2.1 VARIABLE SCREENING

In this case study, the predictive model will be developed for "stress" as a function of "strain", "strain rate", and "temperature". In this book, we have performed

DOI: 10.1201/9781003167372-15

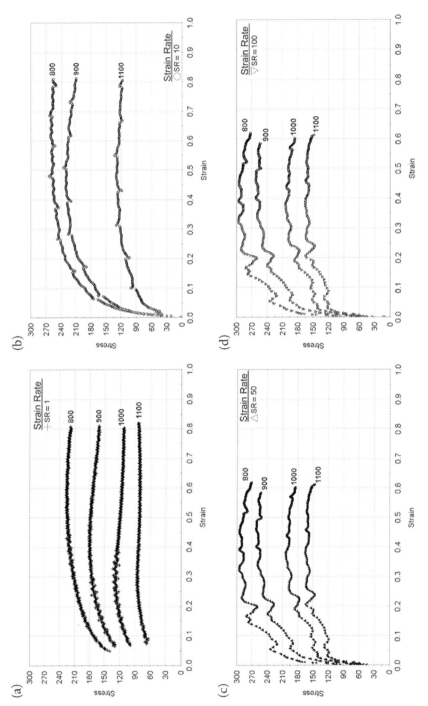

FIGURE 15.1 Stress-strain curve from experimental data at strain rates: (a) 1, (b) 10, (c) 50, and (d) 100 s^{-1}.

TABLE 15.1

Variable Screening through the Smoothing Spline ANOVA Algorithm, along with the Pearson Correlation Coefficient for Stress

Contribution Indices of Variables (ANOVA) Performance Indices (ANOVA)	Stress (ANOVA)	Stress (Pearson Correlation Coefficient)
Strain	0.198	0.275
Strain Rate	0.210	0.329
Temperature	0.592	–0.772
Mean Absolute Error	6.289	
Mean Relative Error	0.052	
Mean Normalized Error	0.021	
R-squared	0.977	

variable screening prior to developing predictive models. Contribution indices for the model inputs were calculated by the smoothing spline ANOVA algorithm. Contribution indices for model parameters as well as model performance indices are shown in Table 15.1.

While working with the supervised machine learning approach using data in tabular form, it is a nice practice to start with statistical analysis. In supervised machine learning, determining prediction accuracy of the models can be done by performing statistical analysis over the actual data and the predicted data. Additionally, a user needs to use different approaches for visualizing the model accuracy. Statistical tools can sometimes show errors that may seem to be large when compared with the actual values, but these models can catch the trends shown in the data. It is important to strike a balance; that is, one must work on lowering error (priority), but they must also give importance to models that can mimic trends observed in the data generated by a system.

15.2.2 Development of Prediction Model for Stress

A significant amount of data was collected through the experiments. Additionally, one set of experiment file was kept aside for using it in the future for checking the prediction accuracy of the developed model. Still, we had a data table of 8,540 rows. This data is well suited for the application of a AI/ML-based algorithm for developing a predictive model for stress. From this table, 90% of the data were assigned to the training set, while 10% of the data were assigned to the testing set.

The workflow has been developed with the modeFRONTIER software. The K-nearest neighbor algorithm was chosen as the supervised machine learning approach for developing models for stress as a function of strain, strain rate, and temperature (Figure 15.2).

The K-NN model was developed from a large amount of experimental data; thus, model prediction accuracy is acceptable. In Figure 15.3, one can observe that the

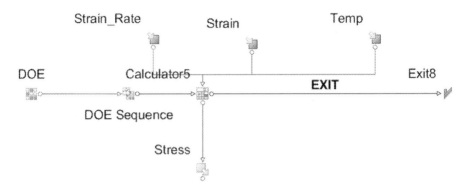

FIGURE 15.2 Workflow developed in modeFRONTIER.

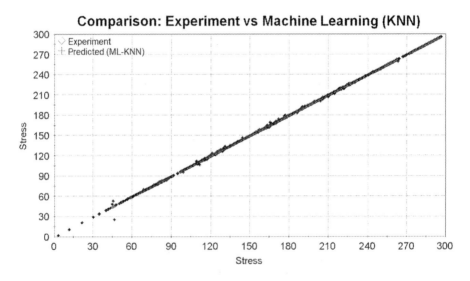

FIGURE 15.3 K-NN model performance over the entire data set.

stress values predicted through K-NN models are close to the 45° line for the entire data set. Ten percent of the data were included in the testing set; thus, the model was not trained over 10% of the data shown in Figure 15.3. Still, predictions are close to the 45° line.

Figure 15.3 has been plotted in a different manner in Figure 15.4. In Figure 15.4, one can observe the temperature and strain rate associated with the experimental data on the figure. In this figure, the fonts are small, but one can easily follow this figure as K-NN model predictions fit the experimental data perfectly.

During model development, one set of files was kept aside. This file was used for checking model prediction accuracy. Figure 15.5 shows the comparison between the actual experimental data and the K-NN model predictions.

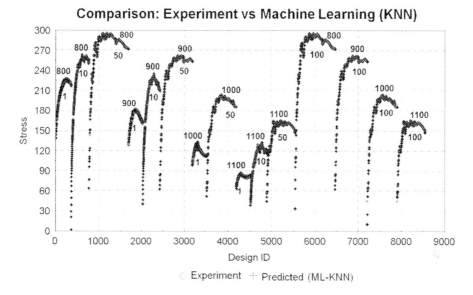

FIGURE 15.4   K-NN model performance over the entire data set showing strain rate and temperature values.

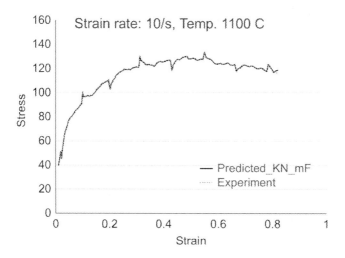

FIGURE 15.5   K-NN model predictions vs. experimental data from a new experiment.

In Figure 15.5, one can observe that machine learning approach can be efficiently used for developing a predictive model for "stress". Additionally, in Figure 15.5, one can observe that these predictive models perform well over new experimental data. K-NN models predicted stress values are in good correlation with the experimental measurements.

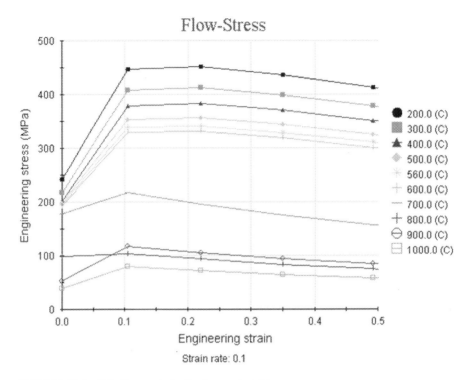

FIGURE 15.6   JMatPro analysis: Strain rate is 0.1/s.

15.3   JMatPRO ANALYSIS AND COMPARISON WITH PRESENT WORK

JMatPro software is widely used in the materials/metallurgy domain for estimating various properties along with a critical phase transformation calculation. In this case study, we will focus on the calculation of flow stress through JMatPro. We have not included information on phase transformations as it is beyond the scope of the current work.

15.3.1   JMatPro ANALYSIS

Figures 15.6–15.8 were obtained from the JMatPro analysis. The composition of a given steel sample was the input for the JMatPro analysis. The strain rate varied as follows: 0.1/s in Figure 15.6, 1/s in Figure 15.7, and 10/s in Figure 15.8.

15.3.2   JMatPro ANALYSIS AND EXPERIMENTAL RESULTS

Figure 15.6 is for a strain rate of 0.1/s, and we do not have experimental data for a strain rate of 0.1/s. In Figures 15.7 and 15.8, the strain value varied from 0 to 0.5. These plots can be compared with Figure 15.1 (experiments). There will be

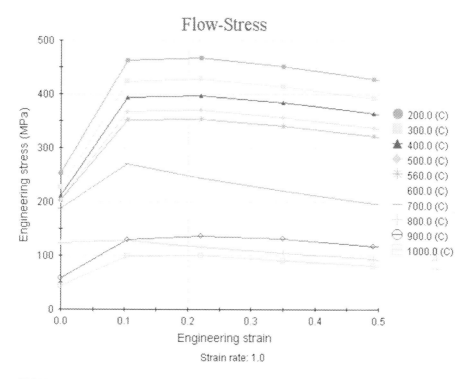

FIGURE 15.7   JMatPro analysis: Strain rate is 1/s.

differences between a simulation software and actual experimental findings. In JMatPro, calculations were performed for specific values of stain, while stress values corresponding to a large set of strain values are generated from the experiments.

JMatPro is one of the most trusted and widely used software for simulating mechanical properties, but it is also a computational platform that follows certain physics. Predictions are made on a set of governing equations. Any CALPHAD-based software utilize databases that are developed through experiments and simulations. Thereafter, these experimental and simulation data are analyzed (fitted) through mathematical and statistical tools. Literature on that material system is used as a reference and for defining a set of governing equations. Thus, there will be a difference in the results obtained from simulation software and experiments. There are lots of factors involved from the experimental side that can be a cause of uncertainty, like uncertainty in determination of composition, temperature, and calibration of the machine, etc. From the software simulation side, in Chapter 1, we presented a case study on uncertainty. In that case study, we discussed uncertainty quantification and propagation. Uncertainty in the determination of thermophysical measurements will affect the models that use that information like enthalpy

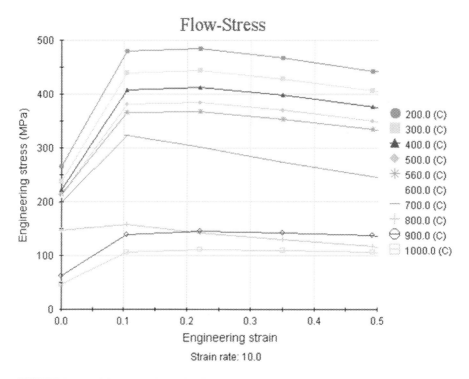

FIGURE 15.8   JMatPro analysis: Strain rate is 10/s.

calculation, calculation of phase diagram, etc. Simulation of mechanical properties
is a complex task and one can expect the difference between experimental and
simulation software predictions. In this book, we have presented case studies where
we have demonstrated ways of calibrating parameters in CALPHAD-based
software.

15.3.3 EXPERIMENTS, JMATPRO ANALYSIS, AND SUPERVISED MACHINE LEARNING

In this work, we have acquired data from actual experiments. We have developed
predictive models through a supervised machine learning algorithm. Predictive
models are in good correlation with experimental findings.

In JMatPro, strain varied from 0 to 0.5 and strain rate varied from 0.1 to 10.0. In
the experimental data in Figure 15.1, strain varies from 0 to about 0.8 for a strain
rate of 1/s and 10/s. For a strain rate of 50/s and 100/s, strain varies from 0 to about
0.6 (Figure 15.1). We used this information as the basis of the work that has been
presented in the next section of this chapter.

Thus, we used the supervised machine learning based predictive model (KNN)
of stress for extrapolation.

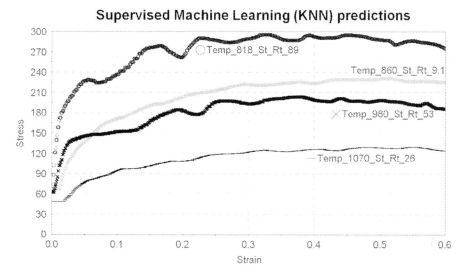

FIGURE 15.9 Supervised Machine Learning (K-NN): Prediction of stress for new experimental parameters.

15.4 DISTRIBUTION OF STRESS STUDIED THROUGH SUPERVISED MACHINE LEARNING MODEL PREDICTIONS

15.4.1 PREDICTION FOR NEW EXPERIMENTAL PARAMETERS

We generated new values for experimental parameters strain, temperature, and strain rate. Then we used the stress model for predicting stress. Figure 15.9 shows the plot where model predictions for new experimental parameters have been shown.

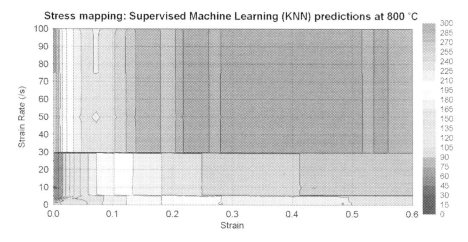

FIGURE 15.10 Stress distribution at 800°C predicted by the supervised ML (K-NN) algorithm.

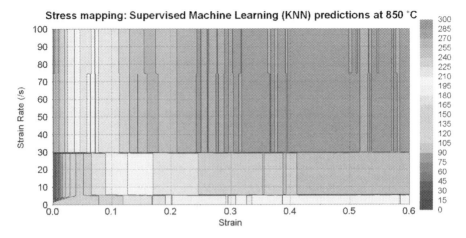

FIGURE 15.11 Stress distribution at 850°C predicted by the supervised ML (K-NN) algorithm.

Figure 15.9 can be compared with Figure 15.1, and one will observe that the model predictions are able to catch the trends shown by experimental data. Thus, we proceeded further for using the predictive model for stress for developing plots for stress distribution for a given temperature.

15.4.2 STRESS MAPPING: SUPERVISED MACHINE LEARNING PREDICTIONS

In the experimental data (Figure 15.1), strain varies from 0 to about 0.8 for a strain rate of 1/s and 10/s. For a strain rate of 50/s and 100/s, strain varies from 0 to about 0.6 (Figure 15.1). In JMatPro, strain varied from 0 to 0.5. Thus, we chose the strain

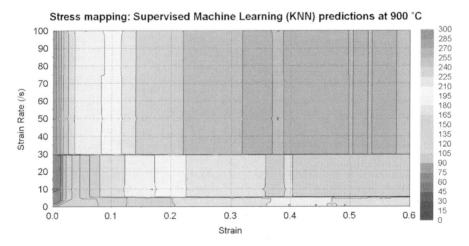

FIGURE 15.12 Stress distribution at 900°C predicted by the supervised ML (K-NN) algorithm.

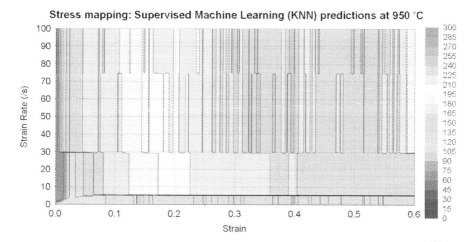

FIGURE 15.13 Stress distribution at 950°C predicted by the supervised ML (K-NN) algorithm.

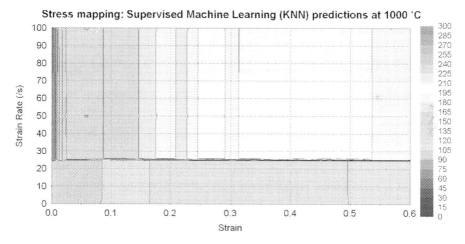

FIGURE 15.14 Stress distribution at 1000°C predicted by the supervised ML (K-NN) algorithm.

value between 0 and 0.6. This will cover the strain range shown for experiments and shown in JMatPro. The strain rate varied between 0 and 100.

Figure 15.10 shows the distribution of stress at 800°C, over the strain and strain rate bounds defined in this section. Similar plots have been shown for 850°C in Figure 15.11, 900°C in Figure 15.12, 950°C in Figure 15.13, 1000°C in Figure 15.14, 1050°C in Figure 15.15, and 1100°C in Figure 15.16.

Figures 15.10 and 15.16 can be used for reference while performing experiments. Simulation software like JMatPro, ANSYS, COMSOL, ABAQUS, etc. are routinely used for the estimation of various mechanical properties. All of these

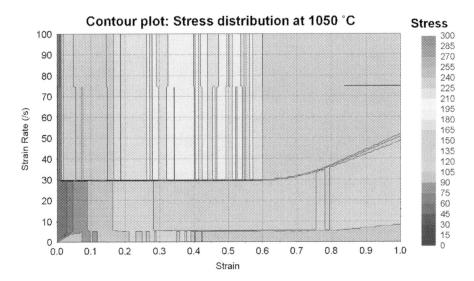

FIGURE 15.15 Stress distribution at 1050°C predicted by the supervised ML (K-NN) algorithm.

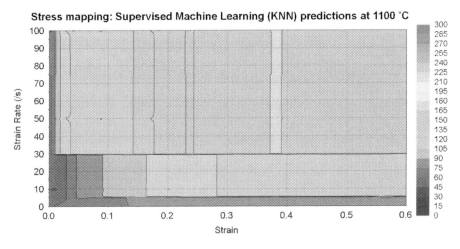

FIGURE 15.16 Stress distribution at 1100°C predicted by the supervised ML (K-NN) algorithm.

software programs require information on material prior to performing simulations. In this work, we have not used any information about the material; still, we have developed a predictive model that is capable of predicting stress for new experimental parameters. Development of the model takes less than two minutes, while prediction of stress for thousands of new parameters can be performed within a minute. These models can be used in other software or computer platforms.

16 Computational Platforms Used in This Work

16.1 COMPUTATIONAL INFRASTRUCTURE

16.1.1 COMPUTER FOR ARTIFICIAL INTELLIGENCE (AI)–BASED WORK

All the work was performed on a laptop. The operating system was Windows 10, Core i7 processor (CPU) with 32 GB RAM. This computer was also used for generating results through the Schrodinger Materials Science Suite. It is better to have HPC access for using this software. Linux distribution (Ubuntu 18.04) was installed in a virtual box on this computer, and it was also used for some calculations.

16.1.2 CALPHAD-BASED WORK

Thermocalc was installed on a desktop computer in a computer lab. The operating system was Windows 10, Core i7 processor (CPU) with 16 GB RAM.

16.1.3 ONLINE PLATFORMS USED FOR GENERATING RESULTS

- **Google Colab:** (Colab 2021)
  It is a Cloud-based environment with free computational resources. A premium version is also available in several countries. It is used for developing DL-ANN models in the Python programming language. Several packages are preinstalled, but a user can install their own package too.
- **CITRINE Informatics**: (O'Mara et al. 2016)
  It is used for developing predictive models through the random forest algorithm and inverse design in the case study in Chapter 7.
- **ABACUS AI:** (Abacus AI 2021)
  It is used for developing the deep learning artificial neural network model for a case study with 100 variables. This case study is presented in this chapter.
- **JupyterHub (sponsored by IBM)**: (JupyterHub 2021).
  This platform was used for generating results presented in a case study in Chapter 1. Access to this platform was provided in a workshop organized by the PyCalphad group in June 2021.

16.1.4 ANDROID PHONE (REDMI NOTE 7 PRO)

An Android phone with 6 GB RAM and octa-core processor (CPU) was also used in this work. Deep learning artificial neural network (DL-ANN) models developed on

DOI: 10.1201/9781003167372-16

the computer can be used on an Android phone. DL-ANN models were developed in the Python programming language. It can be used on an Android phone through the "Pydroid" app available in the Google play store. If one uses only the "Keras" package for developing DL-ANN, then the models can be used for free. If a user uses libraries associated with "TensorFlow", then the models will not work for free, and one has to purchase the premium version.

16.1.5   APPLE IPHONE 6S PLUS

MATLAB® app can be efficiently used on a mobile device, but the mobile version will not allow a user to access the GUI. In Chapters 4 and 12, we developed GUI/ APP in MATLAB. On the phone, one will not be able to use the GUI, but computer codes can be executed. With small sub-routines, one can plot figures and all the models can be analyzed efficiently on an iPhone.

16.1.6   SUPERCOMPUTER/HIGH-PERFORMANCE COMPUTING (HPC)

In Chapter 6, calculations performed through VASP software need HPC access.

HPC resources were used for generating a portion of results for one case study. Apart from this, all the work was performed on a normal computer.

16.2   ESTECO modeFRONTIER

This is a proprietary software used extensively throughout this book. It was used for developing supervised machine learning models through several different approaches. For working with an unsupervised machine learning algorithm, it is one of the best platforms available.

16.3   MATLAB (MATLAB 2019)

MATLAB is a proprietary software. It is one of the best platforms available if one wants to develop a graphical user interface or APP in MATLAB. One can even develop stand-alone software for users without MATLAB access. In Chapters 12 and 4, we developed a GUI/APP in MATLAB. In Chapter 11, most of the calculations for Furnace 3 were performed in MATLAB.

Programs developed in other languages can be used in MATLAB. Regular codes in MATLAB are converted into MEX files for improving the performance of that code in MATLAB. MEX files developed in another computer platform can also be easily used in MATLAB.

16.3.1   DEEP LEARNING TOOLBOX

Deep learning Toolbox was used for developing an artificial neural network model in Chapter 4. A user can even access complex models like AlexNET, GoogleNET, etc. A user can also develop their own models in MATLAB by using several supervised machine learning algorithms.

16.3.2 PlatEMO: Evolutionary Multi-Objective Optimization (Tian et al. 2017)

This toolbox can be used for performing multi-objective optimization in MATLAB. In this work, we have not presented any result using PlatEMO, but it is an efficient toolbox.

16.4 SIGMA TECHNOLOGIES: IOSO (EGOROV 1998)

IOSO, a proprietary software, stands for indirect optimization on the basis of self-organization. The results generated through IOSO algorithms have been used in Chapters 5 and 7. It is an efficient software for solving multi-objective design optimization problems (Egorov-Yegorov and Dulikravich 2005, Jha et al. 2014). A user will need some understanding of the FORTRAN language for using this software.

16.5 CYBERDYNE: KIMEME

Cyberdyne is a proprietary software. It can be used for developing supervised machine learning models. One can perform multi-objective optimization in this software using several classes of evolutionary or genetic algorithms. A user may need some basic understanding of JAVA language for using this software. In Chapter 11 for Furnace 3, this software was used to generate parts of the case study.

16.6 PYTHON PROGRAMMING LANGUAGE PACKAGES

16.6.1 TensorFlow and Keras: Open Source

TensorFlow can be used for designing highly flexible and complex neural network models (Abadi et al. 2016). The Keras package is a user-friendly API for TensorFlow (Chollet 2015). In a few lines of code, one can develop highly complex artificial neural network (ANN) models through Keras. We have developed models using these packages on a laptop computer as well as on the Google Colab platform. As mentioned before, DL-ANN models can be used on an Android phone. A user can find it challenging, but it works.

16.6.2 PyTorch (Paszke et al. 2017)

PyTorch is strictly GPU-based, while TenforFlow has both CPU and GPU versions. PyTorch can be used for developing deep learning models. It is mostly used in image analysis.

16.7 R-STUDIO (R-STUDIO 2021)

R-Studio is an integrated development environment (IDE) for R software. We have used this software mainly for statistical calculations in this book. This software can also be used for developing machine learning models.

16.8   IBM SPSS (IBM SPSS 2015)

IBM SPSS is a proprietary software. SPSS stands for statistical package for the social sciences. It is used for statistical analysis and is widely used by market researchers, health researchers, research firms, government, training researchers, marketing arrangements, and data researchers.

16.9   WEKA: OPEN SOURCE (HALL ET AL. 2009)

WEKA stands for Waikato environment for knowledge analysis (WEKA). We have used it for statistical analysis in several case studies. It can be used for data mining, preprocessing, clustering, classification, regression, visualization, and feature selection. Several machine learning algorithms can be accessed in WEKA. It is based on the Java programming language. At times, a user may need some understanding of Java while working with this software.

16.10   CITRINE INFORMATICS (O'MARA ET AL. 2016)

CITRINE Informatics was used for generating results in Case Study 7. We have listed various steps that need to be followed for developing models. We have also listed steps for using this platform for performing inverse design.

16.11   KNIME (KNIME 2021)

KNIME is quite efficient for performing statistical calculations and developing predictive models based on supervised machine learning. In a few steps, a user can design a simple workflow and perform complex tasks with the click of a few buttons. If a user does not want to write a computer program, this platform will be extremely helpful for them. It has a friendly support team. We have used this platform for several case studies. We will post a few case studies on the KNIME website.

16.12   RAPIDMINER (RAPIDMINER 2021)

RapidMiner can be used for performing complex tasks in a few clicks of a button. A user does not need to write any program for using this software. Complex tasks like data preparation, machine learning, deep learning, and predictive analytics have been performed for a few case studies in this book. Execution time is quite fast.

16.13   ATOMIC SCALE SIMULATION-BASED SOFTWARE

Atomic-scale simulation is performed via various approaches like ab-initio, density functional theory (DFT), molecular dynamics, etc. Atomic-scale simulations are helpful in identifying most stable structures and performing various simulations that provide vital information prior to synthesis of a material. These simulations generate a significant amount of data that can be efficiently utilized by AI-based algorithms. A few types of software used by us are as follows.

16.13.1 MATERIALS PROJECT (JAIN ET AL. 2013)

Materials Project is an open-source platform for obtaining information on various structures prior to performing atomic-based simulations.

16.13.2 VIENNA ATOMIC SIMULATION PACKAGE (VASP) (VASP 2021)

We used VASP for performing calculations for determining the magnetic state of various Fe-Si phases in Chapter 6.

16.13.3 SCHRODINGER MATERIALS SCIENCE SUITE (BOCHEVAROV ET AL. 2013)

The Schrödinger materials science suite is used for performing computational chemistry simulations at the atomic scale. In this work, we used the JAGUAR module of this software. Another prominent application is in the field of drug discovery. It also allows users to utilize AI-based algorithms for accelerating the process of discovery of new structures and optimization of various properties. In this work, we included case study on designing new refrigerants by combining AI-based algorithms' predictions along with atomistic simulations based on computational chemistry.

16.14 CASE STUDY #16: DEVELOPING AI-BASED MODEL FOR 100 VARIABLE PROBLEMS IN ABACUS AI

16.14.1 ABACUS AI: OPEN SOURCE AND PROPRIETARY (ABACUS AI 2021)

ABACUS AI is an AI/ML research start-up. It is a Cloud-based AI platform where a user can develop complex models based on the concepts of deep learning. We have used several concepts of artificial intelligence for developing predictive models in this book. We have developed deep learning artificial neural network (DL-ANN) by writing a computer code in the Python programming language using TensorFlow/Keras libraries. A deep learning algorithm was also used in RapidMiner software. In Chapter 13, we have analyzed the performance of several modules included in mode-FRONTIER and KNIME. In Chapter 13, models were developed for Schittkowski test cases with two and five variables.

In this case, we selected a test problem with 100 variables and developed models in ABACUS AI. Our data set consists of one (1) function, 100 variables, and 2,000 rows. In ABACUS AI, one needs to upload the data, assign the columns for variables and objectives and start the training process. Once the model will be developed, a user will receive an email. The user can then log in to the system and check the model performance.

A user does not need to write a code. The entire process is fully automated and user friendly.

16.14.2 DEEP LEARNING MODEL

Figure 16.1 shows the plot containing models developed in ABACUS AI and their relative performance. The last model in the plot, ABACUS deep learning-best fit neural network, was used for further analysis.

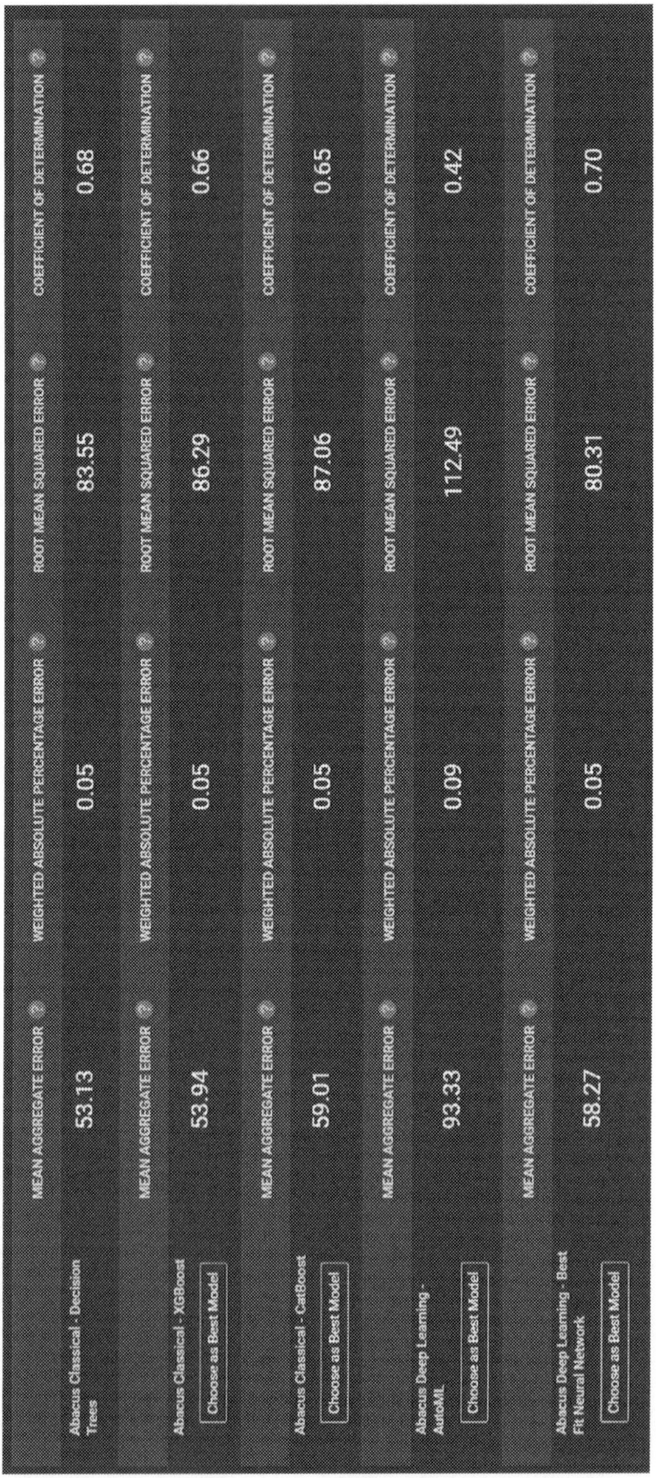

	MEAN AGGREGATE ERROR	WEIGHTED ABSOLUTE PERCENTAGE ERROR	ROOT MEAN SQUARED ERROR	COEFFICIENT OF DETERMINATION
Abacus Classical - Decision Trees	53.13	0.05	83.55	0.68
Abacus Classical - XGBoost (Choose as Best Model)	53.94	0.05	86.29	0.66
Abacus Classical - CatBoost (Choose as Best Model)	59.01	0.05	87.06	0.65
Abacus Deep Learning - AutoML (Choose as Best Model)	93.33	0.09	112.49	0.42
Abacus Deep Learning - Best Fit Neural Network (Choose as Best Model)	58.27	0.05	80.31	0.70

FIGURE 16.1   ABACUS AI: Developed models and error metrics.

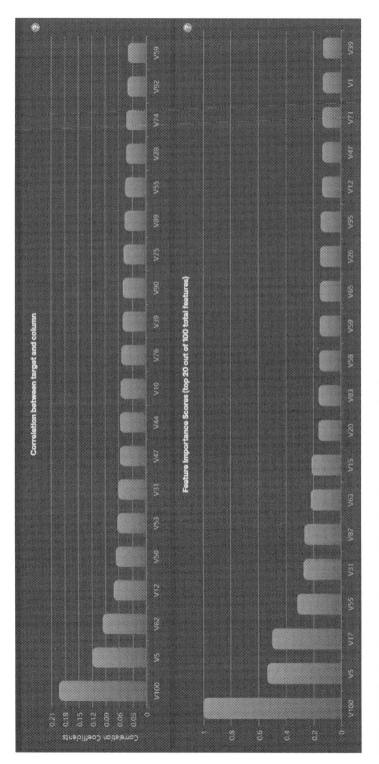

FIGURE 16.2 Variable screening: Top 20 contributors are shown in the figure.

In ABACUS AI, the model performance is analyzed through a parameter termed *weighted absolute percentage error (WAPE)*. WAPE is defined as shown in equation 16.1.

$$WAPE = \frac{\sum_{i=1}^{i=n} |Actual_i - Predicted_i|}{\sum_{i=1}^{i=n} Actual_i}$$ (16.1)

The deep learning model selected in this work had a WAPE value of 0.05.

Throughout this book, we have performed variable screening. Variable screening performed in this work has been reported in Figure 16.2. Since this is a 100-variable problem, the top 20 contributors have been plotted in the figure.

Figure 16.3 shows the plot for comparison between actual and predicted values. In these types of plots, predicted values are expected to be close to the 45-degree line. In Figure 16.3, lots of predicted values are closer to the 45-degree line. A few points are scattered on the figure, but that is expected as we are dealing with a 100-variable problem in this case.

Figure 16.4 shows the window that can be used for checking the prediction capability of the deep learning model developed in ABACUS AI. One can experiment with the test data or use the prediction API. Prediction API will be helpful for deployment of the model for a specific purpose.

ABACUS AI platform can be used for a wide range of problems. In this case study, it was able to accurately fit a 100-variable case study. Usually we rarely get

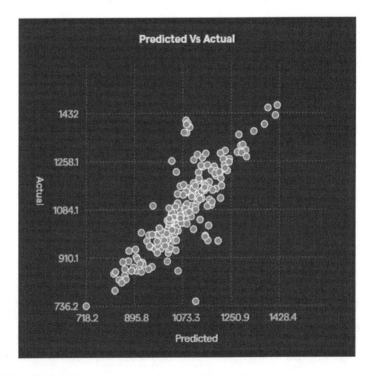

FIGURE 16.3   Model accuracy: Actual vs. predicted value.

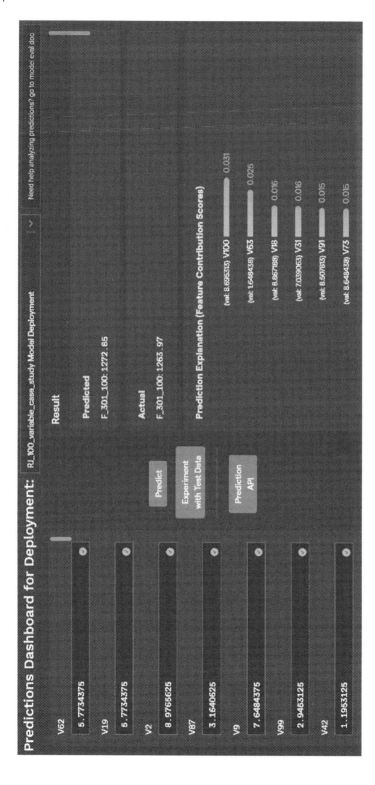

FIGURE 16.4 Model prediction.

such high-dimensional problems to solve. Thus, a user must try this platform for development and deployment of their models.

16.15 CONCLUSIONS

In this chapter, we have listed the details of several computational platforms that were used for developing case studies included in this book.

We have listed the software and programming language used in the case studies. This list includes a number of open-source and proprietary software. Some of the software needs some programming, while for most kinds of software, a user does not need to write a single line of code.

In this book, we have presented several case studies where we have written our own code and we have presented several case studies where a user does not need to write a single line of code.

Thus, it is possible for a user to utilize their data and analyze it through several concepts of artificial intelligence by the click of a few buttons in open-source and proprietary software.

Again, in data analysis, the most important part is having a problem in hand. If a person has a problem, then they can go ahead and arrange relevant data. Thereafter, they can move ahead with the application of artificial intelligence concepts in solving their problem, but, most important is the knowledge of the physics of the problem.

References

Abacus A.I. https://abacus.ai/ (accessed on June 15, 2021).

Abadi, M., Barham, P., and Chen, J. et al. 2016. TensorFlow: A system for large-scale machine learning, 12th USENIX Symposium on Operating Systems Design and Implementation (OSDI 16), USENIX Association, pp. 265–283.

Aggour, K.S., Gupta, V.K., Ruscitto, D., Ajdelsztajn L., Bian, X., et al. 2019. Artificial intelligence/machine learning in manufacturing and inspection: A GE perspective. *MRS Bull*, 44(7), 545–558.

Agren, J. 2015. Nucleation-a challenge in the modelling of phase transformations. International Conference on Solid-Solid Phase Transformations in Inorganic Materials, PTM 2015, Canada, pp. 9–14.

Akaike, H. 1974. A new look at the statistical model identification. *IEEE Transactions on Automatic Control*, 19(6), 716–723.

Aleixo, G.T., Afonso, C., Coelho, A., and Caram, R. 2008. Effects of omega phase on elastic modulus of Ti-Nb alloys as a function of composition and cooling rate. *Solid State Phenom*, 138, 393–398.

Andersen, S.J., Marioara, C.D., Friis, J., Wenner, S., and Holmestad, R. 2018. Precipitates in aluminium alloys. *Advances in Physics: X*, 3(1), 1479984.

Andersson, J.O., Helander, T., Höglund, L., Shi, P.F., and Sundman, B. 2002. Thermo-Calc and DICTRA, Computational tools for materials science. *CALPHAD*, 26, 273–312.

Angstenberger, J. 1996. Artificial neural networks — ICANN 96. *Lecture Notes in Computer Science*, 1112/1996, 203–208.

Arróyave, R., and McDowell, D.L. 2019. Systems approaches to materials design: Past, present, and future. *Annual Review of Materials Research*, 49(1), 103–126.

Assadiki, A., Esin, V.A., Bruno, M., and Martinez, R. 2018. Stabilizing effect of alloying elements on metastable phases in cast aluminum alloys by CALPHAD calculations. *Computational Materials Science*, 145, 1–7.

Babich, A., Formoso, A., Garcia, L., Gudenau, H.W., Mavrommatis, K., and Froehling, C. 2002. Choice of technological regimes of a blast furnace operation with injection of hot reducing gases. *Revista de Metalurgia*, 38(4), 288–305.

Bakshi, S.R., Lahiri, D., Patel, R.R., & Agarwal, A. (2010). Nanoscratch behavior of carbon nanotube reinforced aluminum composite coatings. *Thin Solid Films*, 58, 1703–1711.

Bale, C., Chartrand, P., Degterov, S., Eriksson, G., Hack, K., Mahfoud, R.B., Melançon, J., Pelton, A., and Petersen, S. 2002. Factsage thermochemical software and databases. *CALPHAD*, 26(2), 189–228.

Bale, C.W., Bélisle, E., and Chartrand, P. et al. 2016. FactSage thermochemical software and databases, 2010–2016. *CALPHAD: Computer Coupling of Phase Diagrams and Thermochemistry*, 54, 35–53.

Banks, B.S. 1999. *Neural network based modelling and data mining of blast furnace operations*, Master of Engineering Thesis, Department of Electrical Engineering and Computer Science, Massachusetts Institute of Technology.

Bardel, D., Perez, M., Nelias, D. et al. 2014. Coupled precipitation and yield strength modelling for non-isothermal treatments of a 6061 aluminium alloy. *Acta Mater*, 62, 129–140.

Basheer, I.A., and Hajmeer, M. 2000. Artificial neural networks: Fundamentals, computing, design and application. *Journal of Microbiological Methods*, 43, 3–31.

Bhadeshia, H.K.D.H. 1999. Neural networks in materials science. *ISIJ International*, 39(10), 966–979.

Bhargava, S., Dulikravich, G.S., and Egorov, I.N. 2010. Design of molecules for pareto-optimum functionalities, Inverse Problems, Design and Optimization Symposium, João Pessoa, Brazil, August 25–27.

Biswas, A.K. 1981. *Principles of Blast Furnace Ironmaking*, SBA Publications: Calcutta.

Bochevarov, A.D., Harder, E., Hughes, T.F., Greenwood, J.R., Braden, D.A., Philipp, D.M., Rinaldo, D., Halls, M.D., Zhang, J., and Friesner, R.A. 2013. Jaguar: A high-performance quantum chemistry software program with strengths in life and materials sciences. *International Journal of Quantum Chemistry*, 113(18), 2110–2142.

Bocklund, B., Otis, R., & Egorov, A. (2019). ESPEI for efficient thermodynamic database development, modification, and uncertainty quantification: application to Cu-Mg. *MRS Communications*, 9, 618–627. 10.1557/mrc.2019.59.

Bonvalet, M., Philippe, T., Sauvage, X., and Blavette, D. 2015. Modeling of precipitation kinetics in multicomponent systems: Application to model superalloys. *Acta Mater*, 100, 169–177.

Box, G.E.P., and Draper, N.R. 1987. *Empirical Model-Building and Response Surfaces*, Wiley: New York.

Boyce, B.F., McWilliams, S., Mocan, M.Z., Elder, H.Y., Boyle, I.T., and Junor, B.J.R. 1992. Histological and electron microprobe studies of mineralization in aluminum related osteomalacia. *Journal of Clinical Pathology*, 45, 502–508.

Buschow, K.H.J., and de Boer, F.R. 2003. Soft-magnetic materials. In: *Physics of Magnetism and Magnetic Materials*, Springer: Boston, MA.

Chakraborti, N. 2004. Genetic algorithms in materials design and processing. *International Materials Reviews*, 49(3-4), 246–260.

Chang, K., and Moelans, N. (2015). Phase-field simulations of the interaction between a grain boundary and an evolving second-phase particle. *Philosophical Magazine Letters*, 95, 202–210. 10.1080/09500839.2015.1031845.

Chen, Q., Jeppsson, J., and Agren, J. 2008. Analytical treatment of diffusion during precipitate growth in multicomponent systems. *Acta Mater*, 56, 1890–1896.

Chollet, F. 2015. Keras. https://github.com/keras-team/keras (accessed on June 15, 2021).

Cinkilic, E., Yan, X., and Luo, A. 2020. Modeling precipitation hardening and yield strength in cast Al-Si-Mg-Mn Alloys. *Metals*, 10, 1356.

Coello Coello, C.A., Van Veldhuizen, D.A., and Lamont, G.B. 2002. *Evolutionary Algorithms for Solving Multi-Objective Problems*, Kluwer: New York, NY.

Collet, P. 2007. Genetic programming. In: J.-P. Rennard (Ed.), *Handbook of Research on Nature Inspired Computing for Economics and Management*, Idea: Hershey, PA, pp. 59–73.

Cullity, B., and Graham, C. 2009. Chapter 14. Hard magnetic materials. In: *Introduction to Magnetic Materials*, 2nd ed., Wiley-IEEE Press, New York. pp. 477–504.

Cyber Dyne S.r.l. Kimeme. A new flexible platform for multi-objective and multi-disciplinary optimization. http://www.kimeme.com/ (accessed on 30 May, 2021)

De, S.K., Deva, A., Mukhopadhyay, S., and Jha, B.K. 2011. Assessment of formability of hot-rolled steel through determination of hole-expansion ratio. *Materials and Manufacturing Processes*, 26(1), 37–42.

De, S.K., Deva, A., Mukhopadhyay, S., Jha, B.K., and Chaudhuri, S.K. 2007. Effect of boron addition on microstructure and mechanical properties of low carbon aluminum killed steel. *Steel India*, 29, 61–67.

De Luca, A., Dunand, D.C., and Seidman, D.N. 2018. Scandium-enriched nanoprecipitates in aluminum providing enhanced coarsening and creep resistance. In: O. Martin (Ed.), *Light Metals*, Springer International Publishing: Cham, pp. 1589–1594.

Deb, K. 2001. *Multi-Objective Optimization using Evolutionary Algorithms*, John Wiley and Sons: Chichester, UK.

Deva, A., De, S.K., and Jha, B.K. 2008. Effect of B/N ratio on plastic anisotropy behavior in low carbon aluminum killed steel. *Materials Science and Technology*, 24(1), 124–126.

Deva, A., and Jha, B.K. 2016. *Bwekaoron-Added Low Carbon Unalloyed Steel*, *Encyclopedia of Iron, Steel, and Their Alloys*, CRC Press: Boca Raton, pp. 358–375.

Deva, A., Jha, B.K., and Mishra, N.S. 2011. Silicon as grain refiner in niobium microalloyed hot rolled steel. *Materials Science and Technology*, 27(3), 710–712.

Dilon, H. 2014. Effects of heat treatment and processing modifications on microstructure in AlNiCo-8h permanent magnetic alloys for high temperature applications. Graduate Thesis and Dissertations, Iowa State University. http://lib.dr.iastate.edu/etd/13867

Dimiduk, D.M., Holm, E.A., and Niezgoda, S.R. 2018. Perspectives on the impact of machine learning, deep learning, and artificial intelligence on materials, processes, and structures engineering. *Integrating Materials and Manufacturing Innovation*, 7, 157–172.

Domingo, J.L. 2002. Vanadium and tungsten derivatives as antidiabetic agents. *Biological Trace Element Research*, 88, 97–112.

Dorin, T., Ramajayam, M., Babaniaris, S., Jiang, L., and Langan, T.J. 2019. Precipitation sequence in Al–Mg–Si–Sc–Zr alloys during isochronal aging. *Materialia*, 8, 100437.

Dorin, T., Ramajayam, M., Lamb, J., and Langan, T. 2017. Effect of Sc and Zr additions on the microstructure/strength of Al-Cu binary alloys. *Materials Science & Engineering A: Structural Materials*, 707, 58–64.

Du,Q., Jia, L., Tang, K., and Holmedal, B. 2018. Modelling and experimental validation of microstructure evolution during the cooling stage of homogenization heat treatment of Al–Mg–Si alloys. *Materialia*, 4, 70–80.

Dulikravich, G.S., and Colaço, M.J. 2015. Hybrid optimization algorithms and hybrid response surfaces. In: D. Greiner, B. Galvn, J. Periaux, N. Gauger, K. Giannakoglou, G. Winter (Eds.), *Advances in Evolutionary and Deterministic Methods for Design, Optimization and Control in Engineering and Sciences*, Chapter 2, Springer Verlag: Heidelberg, 19–47.

Dulikravich, G.S., Kumar, A., and Egorov, I.N. 2008. Titanium based alloy chemistry optimization for maximum strength, minimum weight and minimum cost using JMatPro and IOSO software. Proceedings of the TMS Annual Meeting, Materials Informatics: Enabling Integration of Modeling and Experiments in Materials Science, New Orleans, LA, USA, 9–13 March 2008.

Dwivedi, V.S., and Jha, B.K. 2002. Evaluation of activation parameters by Reed-Hill model. *Materials Science and Technology*, 18(2), 134–138.

Egorov, I.N. 1998. Indirect optimization method on the basis of self-organization, Curtin University of Technology, Perth, Australia. *Optimization Techniques and Applications (ICOTA'98)*, 2, 683–691.

Egorov-Yegorov, I.N., and Dulikravich, G.S. 2005. Chemical composition design of superalloys for maximum stress, temperature, and time-to-rupture using self-adapting response surface optimization. *Materials and Manufacturing Processes*, 20(3), 569–590.

Elrod, M.J. 1999. Greenhouse warming potentials from the infrared spectroscopy of atmospheric gases. *Journal of Chemical Education*, 76(12), 1702–1705.

Etminan, M., Highwood, E.J., Laube, J.C., McPheat, R., Marston, G., Shine, K.P., and Smith, K.M. 2014. Infrared absorption spectra, radiative efficiencies, and global warming potentials of newly-detected halogenated compounds: CFC-113a, CFC-112 and HCFC-133a. *Atmosphere*, 5, 473–483.

Fan, M., Liu, Y., Jha, R., Dulikravich, G.S., Schwartz, J., and Koch, C. 2016. Effect of Cu-Ni-rich bridges on the microstructure and magnetic properties of alnico alloys. *IEEE Transactions on Magnetism*, 52(8), 1–10.

Ferrari, A., Paulsen, A., Langenkämper, D., Piorunek, D., Somsen, C., Frenzel, J., Rogal, J., Eggeler, G., and Drautz, R. 2019. Discovery of ω-free high-temperature Ti-Ta-X shape memory alloys from first-principles calculations. *Physical Review Materials*, 3, 103605.

Gao, Y.H., Cao, L.F., Kuang, J., Zhang, J.Y., Liu G., and Sun, J. 2020. Dual effect of Cu on the Al3Sc nanoprecipitate coarsening. *Journal of Materials Science and Technology*, 37, 38–45.

Ghosh, A., and Chatterjee, A. 2015. *Ironmaking and Steelmaking Theory and Practice*, PHI Learning Private Limited: Delhi, India, pp. 285–292.

Ghosh, S., and Dimiduk, D. 2011. *Computational Methods for Microstructure-Property Relations*, Springer: NY.

Giri, B.K., Pettersson, F., Saxén, H., and Chakraborti, N. 2013. Genetic programming evolved through bi-objective genetic algorithms applied to a blast furnace. *Materials and Manufacturing Processes*, 28(7), 776–782.

Gluck, T. 2011. Report on the positive results obtained by the modern surgical experiment regarding the suture and replacement of defects of superior tissue, as well as the utilization of re-absorbable and living tamponade in surgery. *Clinical Orthopaedics and Related Research*, 469, 1528–1535.

Google Colab. https://colab.research.google.com (accessed on June 15, 2021).

Gudkov, S.V., Simakin, A.V., Sevostyanov, M.A., Konushkin, S.V., Losertová, M., Ivannikov, A.Y., Kolmakov, A.G., and Izmailov, A.Y. 2020. Manufacturing and study of mechanical properties, structure and compatibility with biological objects of plates and wire from new Ti-25Nb-13Ta-5Zr Alloy. *Metals*, 10(12), 1–14.

Gupta, R.K., and Pant, B. 2018. Chapter 4—Titanium aluminides. In: Mitra, R. (Ed.), *Intermetallic Matrix Composites*, Woodhead Publishing, Elsevier: Duxford, UK, pp. 71–93.

Gustavsson, J. 2004. Reactions in the lower part of the blast furnace with focus on silicon, Doctoral Thesis, Department of Materials Science and Technology, Division of Applied Process Metallurgy, Royal Institute of Technology, Stockholm Sweeden, pp. 8–11.

Haidemenopoulos, G.N., Katsamas, A.I., and Kamoutsi, H. 2010. Thermodynamics-based computational design of Al-Mg-Sc-Zr alloys. *Metallurgical and Materials Transactions A: Physical Metallurgy and Materials Science*, 41, 888–899.

Hall M., Frank E., Holmes G., Pfahringer B., Reutemann P., and Witten I.H. 2009. The WEKA data mining software: An update. *ACM SIGKDD Explorations Newsletter*, 11(1), 10–18.

Herzer, G. 1993. Nanocrystalline soft magnetic materials. *Physica Scripta*, 307.

Holappa, L. 2019. Historical overview on the development of converter steelmaking from Bessemer to modern practices and future outlook. *Mineral Processing and Extractive Metallurgy*, 128(1-2), 3–16.

Horstemeyer, M.F. 2012. *Integrated Computational Materials Engineering (ICME) for Metals Using Multiscale Modeling to Invigorate Engineering Design with Science*, TMS (The Minerals, Metals & Materials Society), John Willey & Sons: Hoboken, NJ.

IBMSPSS: IBM corp. released 2013. IBM spss statistics for windows, version22.0. armonk,ny:Ibmcorp. http://www-01.ibm.com/software/analytics/spss/ (accessed on March 1, 2015).

Ito, A., Okazaki, Y., Tateishi, T., and Ito, Y. 1995. In vitro biocompatibility, mechanical properties, and corrosion resistance of Ti-Zr-Nb-Ta-Pd and Ti-Sn-Nb-Ta-Pd alloys. *Journal of Biomedical Materials Research*, 29, 893–899.

Jain, A., Ong, S.P., Hautier, G., Chen, W., Richards, W.D., Dacek, S., Cholia, S., Gunter, D., Skinner, D., Ceder, G., & Persson, K.A. 2013. The Materials Project: A materials genome approach to accelerating materials innovation. *APL Materials*, 1, 011002.

Jha, R. 2016. Combined computational-experimental design of high-temperature, high-intensity permanent magnetic alloys with minimal addition of rare-earth elements, Ph.D. Thesis, Manuscript # 3734. https://digitalcommons.fiu.edu/etd/2621/.

Jha, R., & Agarwal, A. 2021. Software (GUI/APP) for Developing AI-Based Models Capable of Predicting Load-Displacement Curve and AFM Image during Nanoindentation. *Coatings*, 11, 299.

Jha, R., Chakraborti, N., Diercks, D.R., Stebner, A.P., and Ciobanu, C.V. 2018. Combined machine learning and CALPHAD approach for discovering processing-structure relationships in soft magnetic alloys. *Computational Materials Science*, 150, 202–211.

Jha, R., Diercks, D.R., Stebner, A.P., Ciobanu, C.V., and Chakraborti, N. 2019. Interfacial energy of copper clusters in Fe-Si-B-Nb-Cu alloys. *Scripta Materialia*, 162, 331–334.

Jha, R., and Dulikravich, G.S. 2019a. Self-organizing maps to design high temperature Ti-Al-Cr-V alloys for maximum thermodynamic stability, Young's modulus and density. *Metals*, 9(5), 537.

Jha, R., and Dulikravich, G.S. 2019b. Design of high temperature Ti–Al–Cr–V alloys for maximum thermodynamic stability using self-organizing maps. *Metals*, 9(5), 537.

Jha, R., and Dulikravich, G.S. 2020. Solidification and heat treatment simulation for aluminum alloys with scandium addition. *Computational Materials Science*, 182, 109749.

Jha, R., and Dulikravich, G.S. 2021. Discovery of new ti-based alloys aimed at avoiding/minimizing formation of α" and ω-phase using CALPHAD and artificial intelligence. *Metals*, 11, 15.

Jha, R., Dulikravich, G.S. et al. 2017. Self-organizing maps for pattern recognition in design of alloys. *Materials and Manufacturing Processes*, 32(10), 1067–1074.

Jha, R., Dulikravich, G.S., Chakraborti, N., Fan, M., Schwartz, J., Koch, C.C., Colaço, M.J., Poloni, C., and Egorov, I.N. 2016. Algorithms for design optimization of chemistry of hard magnetic alloys using experimental data. *Journal of Alloys and Compounds*, 682, 454–467.

Jha, R., Dulikravich, G.S., Colaço, M.J., Fan, M., Schwartz, J., and Koch, C.C. 2017. Magnetic alloys design using multi-objective optimization. In: A. Oechsner, L.M. da Silva, and H. Altenbac (Eds.), *Properties and Characterization of Modern Materials*, Vol. 33, Advanced Structured Materials Series, Springer: Germany, pp. 261–284.

Jha, R., Pettersson, F., Dulikravich, G.S., Saxen, H., and Chakraborti, N. 2015. Evolutionary design of nickel-based superalloys using data-driven genetic algorithms and related strategies. *Materials and Manufacturing Processes*, 30(4), 488–510.

Jha, R., Sen, P.K., and Chakraborti, N. 2014. Multi-objective genetic algorithms and genetic programming models for minimizing input carbon rates in a blast furnace compared with a conventional analytic approach. *Steel Research International*, 85(2), 219–232.

JMatPro 2021. https://www.sentesoftware.co.uk/jmatpro, accessed on March 30 2021.

JupyterHub. https://jupyter.org/hub.

Kalil, T., and Wadia, C. 2011. Materials genome initiative: A renaissance of American manufacturing. https://obamawhitehouse.archives.gov/blog/2011/06/24/materials-genome-initiative-renaissance-american-manufacturing.

Kampmann, R., Eckerlebe, H., and Wagner, R. 2000. Precipitation kinetics in metastable solid solutions-theoretical considerations and application to cu-ti alloys. *Materials Research Society Symposium Proceedings*, 57, 525–542.

Kennedy, J., and Eberhart, R. 1995. Particle swarm optimization. Proceedings of IEEE International Conference on Neural Networks, pp. 1942–1948.

Keras: The Python deep learning library. https://keras.io/ (accessed on December 30, 2020).

Kirkpatrick, S., Gelatt Jr, C.D., and Vecchi, M.P. 1983. Optimization by simulated annealing. *Science*, 220(4598), 671–680.

KNIME. https://www.knime.com/ (accessed on June 15, 2021).

Krasznai, E.Á., Boda, P., Csercsa, A., Ficsór, M., and Várbíró, G. 2016. Use of self-organizing maps in modelling the distribution patterns of gammarids (Crustacea: Amphipoda). *Ecological Informatics*, 31, 39–48.

Kumar, S., and Padture, N.P. 2018. Chapter 5: Materials in the aircraft industry. In: B. Kaufman and C.L. Briant (Eds.), *Metallurgical Design and Industry: Prehistory to the Space Age*, Springer, Cham, pp. 271–346.

Lahiri, D., Singh, V., Keshri, A.K., Seal, S., andAgarwal, A. 2010. Carbon nanotube toughened hydroxyapatite by spark plasma sintering: Microstructural evolution and multi-scale tribological properties. Carbon, 48, 3103–3120. 10.1016/j.carbon.2010.04.047.

Langer, J., and Schwartz, K. 1980. Kinetics of nucleation in near-critical fluids. *Physical Review A*, 21, 948.

Lecun, Y., Bengio, Y., and Hinton, G. 2015. Deep learning. *Nature*, 521(7553), 436–444.

Li, S., Kattner, U.R., and Campbell, C.E. 2017. A computational framework for material design. *Integrating Materials and Manufacturing Innovation*, 6, 229–248.

Liu, Haiyan, Wei, Yueguang, Liang, Lihong, Wang, Yingbiao, Song, Jingru, Long, Hao, & Liu, Yanwei (2020). Microstructure Observation and Nanoindentation Size Effect Characterization for Micron-/Nano-Grain TBCs. Coatings, 10, 345.

Long, M., and Rack, H. 1998. Titanium alloys in total joint replacement—A materials science perspective. *Biomaterials*, 19, 1621–1639.

Lytvynyuk, Y., Schenk, J., Hiebler, M., and Sormann, A. 2014. Thermodynamic and kinetic model of the converter steelmaking process. Part 1: The description of the BOF model. *Steel Research International*, 85, 537–543.

Mahanta, B.K., Jha, R., and Chakraborti, N. 2022. Data-driven optimization of blast furnace iron making process using evolutionary deep learning. In: S. Datta and J.P. Davim (Eds.), *Machine Learning in Industry. Management and Industrial Engineering*, Springer, Cham.

Mantri, S., Choudhuri, D., Behera, A., Hendrickson, M., Alam, T., and Banerjee, R. 2019. Role of isothermal omega phase precipitation on the mechanical behavior of a Ti-Mo-Al-Nb alloy. *Materials Science and Engineering: A*, 767, 138397.

Marker, C. 2017. Development of a knowledge base of Ti-Alloys from first-principles and thermodynamic modeling. Ph.D. Thesis, The Pennsylvania State University, State College, PA, USA, August 2017. https://www.proquest.com/docview/1988756108 (accessed on March 30, 2021).

Markowitz, S. 2009. *The Advanced Materials Revolution: Technology and Economic Change in the Age of Globalization*, John Wiley and Sons, Inc: New York.

MATLAB and Deep Learning Toolbox™ (Formerly Neural Network Toolbox™) Release 2019b. The MathWorks, Inc.: Natick, MA, USA, 2019.

MATLAB and Statistics Toolbox Release 2019b. The MathWorks, Inc.: Natick, MA, USA.

Mcguiness, P., Akdogan, O., Asali, A., et al. 2015. Replacement and original magnet engineering options (ROMEOs): A European seventh framework project to develop advanced permanent magnets without, or with reduced use of, critical raw materials. *JOM*, 67, 1306–1317.

Metropolis, N., Rosenbluth, N., Rosenbluth, M., Teller, A., and Teller, E. 1953. Equation of state calculations by fast computing machines. *Journal of Chemical Physics*, 21, 1087–1092.

Miettinen, K. 1999. *Nonlinear Multiobjective Optimization*, Kluwer: Boston, MA.

modeFRONTIER. http://www.esteco.com/home/mode_frontier.html (accessed on June 15, 2021).

Mohammed, M.T., Khan, Z.A., and Siddiquee, A.N. 2014. Beta titanium alloys: The lowest elastic modulus for biomedical applications: A review. *International Journal of Chemical, Molecular, Nuclear, Materials and Metallurgical Engineering*, 8, 726–731.

Momma, K., and Izumi, F. 2011. VESTA 3 for three-dimensional visualization of crystal, volumetric and morphology data. *Journal of Applied Crystallography*, 44, 1272–1276.

Moelans, N et al. 2015. A phase-field simulation study of irregular grain boundary migration during recrystallization. 89, 012037.

Mondol, S., Alam, T., Banerjee, R., Kumar, S., and Chattopadhyay, K. 2017. Development of a high temperature high strength Al alloy by addition of small amounts of Sc and Mg to 2219 alloy. *Materials Science & Engineering A: Structural Materials: Properties, Microstructure and Processing*, 687, 221–231.

Morgan, D., and Jacobs, R. 2020. Opportunities and challenges for machine learning in materials science. *Annual Review of Materials Research*, 50(1), 71–103.

Mueller, T., Kusne, A.G., and Ramprasad, R. 2016. Machine learning in materials science: Recent progress and emerging applications. *Reviews in Computational Chemistry*, 29, 186–273.

Oliver, W.C., and Pharr, G.M. 1992. An improved technique for determining hardness and elastic modulus using load and displacement sensing indentation experiments. *Journal of Materials Research*. 7, 1564–1583. 10.1557/JMR.1992.1564.

O'Mara, J., Meredig, B., and Michel, K. 2016. Materials data infrastructure: A case study of the citrination platform to examine data import, storage, and access. *JOM*, 68(8), 2031–2034.

Otis, R.A., and Liu, ZK 2017. High-Throughput Thermodynamic Modeling and Uncertainty Quantification for ICME. *JOM*, 69, 886–892.

Otis, R., Bocklund, B., & Liu, Z. (2021). Sensitivity estimation for calculated phase equilibria. *Journal of Materials Research*, 36, 140–150.

Panchal, J.H., Kalidindi, S.R., and McDowell, D.L. 2013. Key computational modeling issues in integrated computational materials engineering. *Computer-Aided Design*, 45(1), 4–25.

Paszke, A., Gross, S., Chintala, S., Chanan, G., Yang, E., et al. 2017. Automatic differentiation in PyTorch. 31st Conference on Neural Information Processing Systems.

Paszkowicz, W. 2013. Genetic algorithms, a nature-inspired tool: A survey of applications in materials science and related fields: Part II. *Materials and Manufacturing Processes*, 28(7), 708–725.

Palasyuk, A., Blomberg, E., & Prozorov, R. 2013. Advances in Characterization of Non-Rare-Earth Permanent Magnets: Exploring Commercial Alnico Grades 5-7 and 9, 65, 862–869.

Pedregosa F., and Varoquaux, G. 2011. Scikit-learn: *Machine learning in Python*, 12, 2825–2830.

Pena, M., Barbakh, W., and Fyfe, C. 2008. Topology-preserving mappings for data visualisation. In: A.N. Gorban, B. Kégl, D.C. Wunsch, and A.Y. Zinovyev (Eds.), *Principal Manifolds for Data Visualization and Dimension Reduction*, Vol. 58, Lecture Notes In Computational Science and Engineering, Springer Berlin Heidelberg: Heidelberg, pp. 131–150.

Penz, F.M., Bundschuh, P., Schenk, J., Panhofer, H., Pastucha, K., and Paul, A. 2017. Effect of scrap composition on the thermodynamics of kinetic modelling of BOF converter. Proceedings of the 2nd VDEh-ISIJ-JK Symposium, Stockholm, Sweden, 12–13 June 2017, pp. 124–135.

Penz, F.M., Schenk, J., Ammer, R., Klösch, G., and Krzysztof, P. 2018. Dissolution of scrap in hot metal under Linz–Donawitz (LD) steelmaking conditions. *Metals*, 8(12), 1078.

Perez, M., Dumont, M., and Acevedo-Reyes, D. 2008. Implementation of classical nucleation and growth theories for precipitation. *Acta Mater*, 56, 2119–2132.

Périgo, E.A., Weidenfeller, B., Kollár, P., and Füzer, J. 2018. Past, present, and future of soft magnetic composites. *Applied Physics Reviews*, 5, 031301.

Pettersson, F., Chakraborti, N., and Saxén, H. 2007. A genetic algorithms based multi-objective neural net applied to noisy blast furnace data. *Applied Soft Computing*, 7(1), 387–397.

Pitchuka, S.B., Lahiri, D., Sundararajan, G., & Agarwal, A. 2014. Scratch induced deformation behavior of cold sprayed aluminum amorphous/nanocrystalline coatings at multiple load scales. J. Therm. Spray Technol., 23, 502–513. 10.1007/s11666-013-0021-x.

Pitchuka, S.B., Jha, R., Guzman, M., Sundararajan, G., & Agarwal, A. 2016. Indentation creep behavior of cold sprayed aluminum amorphous/nano-crystalline coatings. Mater. Sci. Eng. A, 658, 415–421.

Poli, R., Langdon, W.B., and McPhee, N.F. 2008. *A Field Guide to Genetic Programming*, NC Lulu Press: Morrisville.

Pollock, T.M., and Tin, S. 2006. Nickel-based superalloys for advanced turbine engines: Chemistry, microstructure and properties. *Journal of Propulsion and Power*, 22, 361–374.

Polmear, I., John, D., Nie, J.-F., and Qian, M. (Eds.), 2017. Chapter 7—Titanium alloys. In: *Light Alloys*, 5th ed., Butterworth-Heinemann, Elsevier: Oxford, UK, pp. 369–460.

Ramprasad, R., Batra, R., Pilania, G., Mannodi-Kanakkithodi, A., and Kim, C. 2017. Machine learning and materials informatics: Recent applications and prospects. *npj Computational Materials*, 3(54), 1–13.

RapidMiner|Best Data Science & Machine Learning Platform. https://rapidminer.com/ (accessed on June 15, 2021).

Rechenberg, I. 1971. Evolutionsstrategie – Optimierung technischer Systeme nach Prinzipien der biologischen Evolution, Ph.D. Thesis. Reprinted by Fromman-Holzboog (1973).

Rickman, J.M., Lookman, T., and Kalinin, S.V. 2019. Materials informatics: From the atomic-level to the continuum. *Acta Mater*, 168, 473–510.

Rougier, L., Jacot, A., Gandin, C.-A. et al. 2013. Numerical simulation of precipitation in multicomponent Ni-base alloys. *Acta Mater,* 61, 6396–6405.

Røyset, J., and Ryum, N. 2005. Scandium in aluminium alloys. *International Materials Reviews*, 50(1), 19–44.

R-Studio, The R Project for Statistical Computing. https://www.r-project.org/ (accessed on June 15, 2021).

Russell, K.C., and Yamauchi, K. 1980. Nucleation in solids: The induction and steady state effects. *Advances in Colloid and Interface Science*, 13, 205–318.

Sarafoglou, P.I., Serafeim, A., Fanikos, I.A., Aristeidakis, J.S., and Haidemenopoulos, G.N. 2019. Modeling of microsegregation and homogenization of 6xxx Al-alloys including precipitation and strengthening during homogenization cooling. *Materials*, 12, 1421.

Schmitz, G.J., and Prahl, U. 2014. ICMEg – The Integrated Computational Materials Engineering expert group – a new European coordination action. *Integrating Materials*, 3, 20–24.

Schmöle, P., and Lüngen, H.B. 2004. Hot metal production in the blast furnace from an ecological point of view, Presented at the 2nd International Meeting on Ironmaking and 1st International Symposium on Iron Ore, Vitoria, Brazil.

Schrodinger drug discovery. https://www.schrodinger.com/drug-discovery (accessed on June 15, 2021).

Schulz, E., Speekenbrink, M., and Krause, A. 2018. A tutorial on Gaussian process regression: Modelling, exploring, and exploiting functions. *Journal of Mathematical Psychology*, 85, 1–16.

Schwefel, H.-P. 1974. Numerische Optimierung von Computer-Modellen, Ph.D. Thesis. Reprinted by Birkhäuser (1977).

Smith, Martin P. 2017. Blast furnace ironmaking – A view on future developments. *Procedia Engineering*, 174, 19–28.

Sobol, I.M. 1967. Distribution of points in a cube and approximate evaluation of integrals. *USSR*, 7, 86–112.

SPM Modi. 2021 SPM modi and measurement methods, https://www.dme-spm.com/spmmodi.html, accessed on April 30, 2021.

Storn, R., & Price, K. 1997. Differential Evolution - A Simple and Efficient Heuristic for global Optimization over Continuous Spaces. *Journal of Global Optimization*, 11, 341–359. 10.1023/A:1008202821328.

The HDF Group. https://www.hdfgroup.org/about-us/ (accessed on June 15, 2021).

Tancret, F. 2012. Computational thermodynamics and genetic algorithms to design affordable γ'-strengthened nickel–iron based superalloys. *Modelling and Simulation in Materials Science and Engineering*, 20(4), 1–6.

Tancret, F. 2013. Computational thermodynamics, Gaussian processes and genetic algorithms: Combined tools to design new alloys. *Modelling and Simulation in Materials Science and Engineering*, 21, 1–9.

Tang, K., Dua, Q., and Li, Y. 2018. Modelling microstructure evolution during casting, homogenization and ageing heat treatment of Al-Mg-Si-Cu-Fe-Mn alloys. *CALPHAD*, 63, 164–184.

TensorBoard: TensorFlow's visualization toolkit. https://www.tensorflow.org/tensorboard (accessed on June 15, 2021).

Thermo-Calc Software MOBAL4: TCS Al-alloys Mobility Database, v4 (accessed on October 30, 2018).

Thermo-Calc Software TCAL5: TCS Aluminium-based Alloys Database v.5. https://www.thermocalc.com/media/56675/TCAL5-1_extended_info.pdf (accessed October 30, 2018).

Tian, Y., Cheng, R., Zhang, X., and Jin, Y. 2017. PlatEMO: A MATLAB platform for evolutionary multi-objective optimization [Educational Forum]. *IEEE Computational Intelligence Magazine*, 12(4), 73–87.

Trenkler, H.A. 1960. New developments in LD steelmaking. *JOM*, 12, 538–541.

Tsepelev, V.S., and Starodubtsev, Y.N. 2021. Nanocrystalline soft magnetic iron-based materials from liquid state to ready product. *Nanomaterials*, 11(1), 108.

Turkdogan, E.T. 1996. *Fundamentals of Steelmaking*, The Institute of Materials: London, UK, pp. 209–244.

VASP. 2021. https://www.vasp.at/ (accessed on June 15, 2021).

Vanherpe, L., Moelans, N., Blanpain, B., & Vandewalle, S. 2007. Three-dimensional phase field simulations of grain growth in materials containing finely dispersed second-phase particles. 7, 2020001–2020002.

Virtual Materials Design. 2021. https://www.frontiersin.org/research-topics/22754/virtual-materials-design#overview (accessed on July 15, 2021).

Wagner, R., and Kampmann, R. 1991. Homogeneous second phase precipitation. In: R.W. Cahn, P. Haasen, E.J. Kramer (Eds.), *Phase Transformations in Materials*, Materials Science and Technology, Vol. 5, VCH: Weinhein, Germany.

Wu, X.-H. 2006. Review of alloy and process development of TiAl alloys. *Intermetallics*, 14, 1114–1122.

Zhang, C., Jiang, X., Zhang, R. et al. 2019. High-throughput thermodynamic calculations of phase equilibria in solidified 6016 Al-alloys. *Computational Materials Science*, 167, 19–24.

Zhou, P., Song, H., Wang, H., and Chai, T. 2017. Data-driven nonlinear subspace modeling for prediction and control of molten iron quality indices in blast furnace ironmaking. *IEEE Transactions on Control Systems Technology*, 25(5), 1761–1774.

Zhou, C., Tang, G., Wang, J. et al. 2016. Comprehensive numerical modeling of the blast furnace ironmaking process. *JOM*, 68, 1353–1362.

Zitzler, E., Laumanns, M., and Thiele, L. 2001. SPEA2: Improving the performance of the strength pareto evolutionary algorithm, Technical Report 103, Computer Engineering and Communication Networks Lab (TIK), Swiss Federal Institute of Technology (ETH) Zurich.

Index

331

Printed in the United States
by Baker & Taylor Publisher Services